SpringerBriefs in Applied Sciences and Technology

Series Editor

SpringerBriefs present concise summaries of cutting-edge research and practical applications across a wide spectrum of fields. Featuring compact volumes of 50 to 125 pages, the series covers a range of content from professional to academic.

Typical publications can be:

- A timely report of state-of-the art methods
- An introduction to or a manual for the application of mathematical or computer techniques
- A bridge between new research results, as published in journal articles
- A snapshot of a hot or emerging topic
- An in-depth case study
- A presentation of core concepts that students must understand in order to make independent contributions

SpringerBriefs are characterized by fast, global electronic dissemination, standard publishing contracts, standardized manuscript preparation and formatting guidelines, and expedited production schedules.

On the one hand, **SpringerBriefs in Applied Sciences and Technology** are devoted to the publication of fundamentals and applications within the different classical engineering disciplines as well as in interdisciplinary fields that recently emerged between these areas. On the other hand, as the boundary separating fundamental research and applied technology is more and more dissolving, this series is particularly open to trans-disciplinary topics between fundamental science and engineering.

Indexed by EI-Compendex, SCOPUS and Springerlink.

More information about this series at http://www.springer.com/series/8884

Kiran Kumar Poloju

Advanced Materials and Sustainability in Civil Engineering

Kiran Kumar Poloju
Middle East College
Muscat, Oman

ISSN 2191-530X ISSN 2191-5318 (electronic)
SpringerBriefs in Applied Sciences and Technology
ISBN 978-981-16-5948-5 ISBN 978-981-16-5949-2 (eBook)
https://doi.org/10.1007/978-981-16-5949-2

This Springer imprint is published by the registered company Springer Nature Singapore Pte Ltd.
The registered company address is: 152 Beach Road, #21-01/04 Gateway East, Singapore 189721, Singapore

Acknowledgements

I would like to dedicate this book to my family for their tremendous motivation and moral support to complete this book. I wish to express my sincere gratitude to my colleagues, Department of Civil Engineering, the Management of Middle East College, and Dr. Priya Mathew.

Contents

About the Author

Mr. Kiran Kumar Poloju is a well-qualified and resourceful academician with decent experience in the field of civil engineering with a good number of research publications in reputed journals, has approved funding projects, and possesses knowledge on flipped methods of teaching. He has achieved the status of fellowship (FHEA) in recognition of attainment against the UK Professional Standards Framework for teaching and learning support in higher education. He has obtained Chartered Engineer Certification, Life Member and memberships from different professional bodies such as the Institute of Engineers, India, Institution of Civil Engineers, UK, Indian Institution of Bridge Engineers and Indian Green Building Council, India. His research interest includes civil engineering materials, material characterization of geopolymer concrete, recycling landfills industry waste disposals, concrete technology, teaching and learning pedagogies, sustainable and green technology, and artificial neural network applications.

List of Figures

List of Tables

Chapter 1
Concrete

1.1 Introduction to Concrete

Concrete has been proven to be a leading construction material for more than a century. It is obtained by mixing fine aggregates, coarse aggregates, and cement with water and sometimes admixtures in the required proportions. Concrete is the most abundant manmade material in the world. In most developing countries, concrete is the second most utilized resource after water. Generally, concrete is strong in compression and weak in tension. To stabilize concrete strength in tension, reinforcement is provided during the construction process.

1.2 Constituents of Concrete

The ingredients that are required to develop conventional concrete or normal concrete are ordinary Portland cement, fine aggregate, coarse aggregate, and water. All the ingredients used for comprising concrete have their characteristics and play a major role in contributing to the strength properties of concrete.

1.2.1 Importance of the Ingredients in Concrete

Each ingredient of concrete has its properties as stated above. Cement has binding properties which enable it to facilitate bonding fine and coarse aggregates. Cement is a major constituent of concrete since 70–75% of the mass of concrete is occupied by aggregates. The compounds present in cement are responsible for developing the bonding between the aggregates and providing hydration. Water is used as a mixing agent for cement and aggregates during the concrete mixing process to start

K. K. Poloju, *Advanced Materials and Sustainability in Civil Engineering*,
SpringerBriefs in Applied Sciences and Technology,
https://doi.org/10.1007/978-981-16-5949-2_1

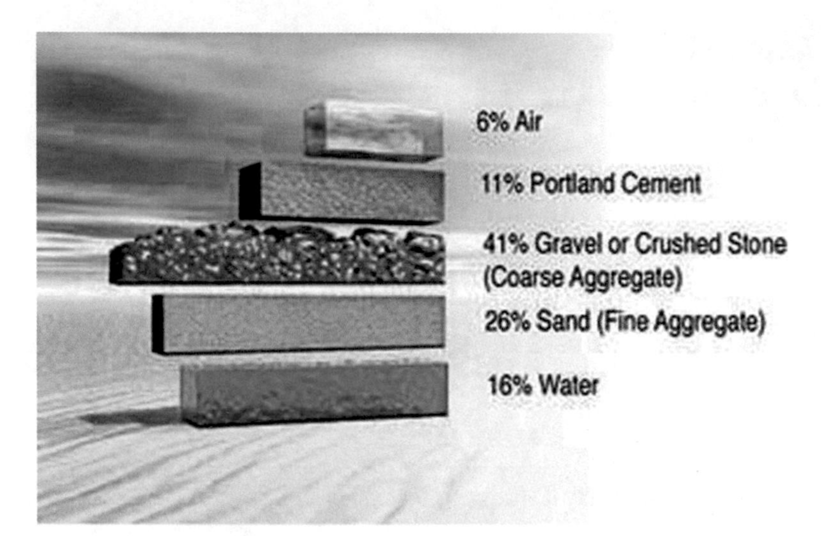

Fig. 1.1 Constituents of concrete

up the hydration process and formation of CSH gel. Generally, when water is added to cement, the hydration process takes place but is not completed; a minimum of 28 days is required to complete the hydration process.

Concrete is fundamentally a combination of cement, aggregates, and water. The cement paste is made up of cement with adequate water content and mixed with aggregates to develop concrete. Ordinary Portland cement consists of around 10–15% of concrete, 60–75% aggregate, and 10–20% water with 5–8% air. The water reacts with the cement chemically for hydration to take place. Concrete is utilized in almost all construction works such as pipes, design structures, foundations, and bridges. The constituents of concrete are shown in Fig. 1.1.

1.3 Characteristics of Concrete

Concrete is a mixture of cement, fine aggregate, coarse aggregate, and water. The tests to be conducted during the two stages of concrete mixing are important to determine the quality of concrete. They are as follows:

(1) Fresh concrete
(2) Hardened concrete.

1.3.1 Fresh Concrete

Fresh concrete is defined as the concrete prepared before the test without any loss due to segregation or bleeding. The concrete mix should be prepared with the appropriate proportion of ingredients such as cement and aggregates with water to maintain the cohesive nature of the mix. This can enable the mix to be poured into the required mould shape and size. The water content in concrete plays a crucial role in developing its cohesive property.

1.3.2 Hardened Concrete

After casting the moulds in the required size and shape, the moulds are placed aside for 24 h for demoulding. The concrete is termed hardened concrete when it is released from moulds and before it is placed in a water tank for curing. The concrete should be hardened and cured for at least 3 days before conducting any tests to determine the quality of concrete. Generally, the strength test is done after 1 day, 3 days, 7 days, 14 days, 21 days, and 28 days after water curing as per the requirement. The compressive strength of the concrete cube specimen being tested after 7 days of water curing will obtain 67% of the full strength of concrete, whereas 90% of strength would be obtained for the concrete cube specimen tested after 14 days of water curing. Thus, to attain the full strength of concrete, it is critical to complete the hydration process for 28 days.

1.3.3 Workability of Concrete

The workability of concrete is defined as the ease with which concrete can be prepared, transported, and stored on-site without losing its homogeneity. The workability of concrete depends on various aspects such as water content, shape and size of the aggregate, level of hydration process, and the use of admixtures. The increase in the water content increases the workability of concrete and vice versa. Generally, slump cone tests, flow table tests, and compaction tests are conducted to determine the workability of fresh concrete. The workability of fresh concrete is directly proportional to the grading of concrete and inversely proportional to the water content/water–cement ratio.

1.3.4 Stability of Concrete

Concrete is poured into the required moulds to determine required tests after completion of the required water curing. If low-quality materials are used to mix concrete, then it is reflected during the process of casting the cubes, as the moulds are required to undergo manual or table vibration to remove the voids present in the concrete to achieve good strength. Sometimes, the concrete might undergo segregation which eventually leads to loss of homogeneity. Therefore, the stability of concrete depends on the use of good quality materials in the right proportion. In other words, the term stability is defined as the ability of concrete to retain its homogeneity after being cast into the moulds (during and after vibration).

1.3.5 Segregation of Concrete

The separation of the ingredients of concrete such as fine aggregate and coarse aggregate from cement is called segregation. Segregation leads to a reduction in the weight and strength of concrete. Segregation may be caused to the use of low-quality ingredients in the concrete mix and increased water content.

1.3.6 Bleeding of Concrete

The bleeding of concrete is the excess of water in the mix. This might be due to the use of more water content during the concrete mix. Bleeding leads to the generation of cracks on hardened concrete due to the insufficient amount of water present in concrete.

1.3.7 Curing of Concrete

The curing of concrete is very important as it retains the moisture content in the concrete to avoid cracks and continue the hydration process to attain full strength. The curing of concrete is generally for 28 days. However, the strength of concrete directly depends on the number of days it is water cured. Decreasing the duration of the water curing process leads to the formation of cracks and instability and a decrease in strength.

1.4 Cement

Cement was first invented by John Smeaton in 1756 by mixing pebbles and powdered brick. Later, Portland cement is manufactured by Joseph Aspidin in 1824 in Portland in England.

Cement is the most important constituent of concrete which assists the bonding between aggregates to achieve maximum strength. The quality of concrete depends on the chemical composition of cement. The chemical composition of cement is shown in Table 1.1.

Ordinary Portland concrete is made of calcium silicate and aluminate. It is prepared by mixing the required amount of limestone, clay, and various minerals in little amounts under a high temperature of around 1500 °C to form a clinker. The clinker is then ground with a little amount of gypsum to deliver a fine powder called ordinary Portland concrete (OPC). At the point when water is blended in with the aggregates, it consolidates gradually with the water to shape a hard mass called concrete. Concrete is a material that absorbs moisture and undergoes hydration when it reacts with cement. Along these lines, the cement remains in better condition as long as it does not interact with moisture. If the cement is over three years of age, its quality needs to be tested before use. The Bureau of Indian Standards (BIS) has arranged OPC into three distinctive grades. These grades are based on the compressive strength of concrete mortar (Cement and sand with water) accomplished following 28 days of water curing. The different grades are as follows:

1. 33 grade
2. 43 grade
3. 53 grade.

33 Grade of cement: This grade of cement achieves a compression strength of 33 N/mm^2 after 28 days.

43 Grade of cement: This grade of cement achieves a compression strength of 43 N/mm^2 after 28 days.

Table 1.1 Chemical composition of cement

S. No.	Material	Composition (%)
1	Lime (CaO)	60–67
2	Silica (SiO_2)	17–25
3	Alumina (Al_2O_3)	3–8
4	Calcium sulphate ($CaSO_4$)	3–4
5	Iron oxide (Fe_2O_3)	3–4
6	Magnesia (MgO)	0.5–4
7	Sulphur trioxide (SO_3)	1–2
8	Alkalis	< 1

53 Grade of cement: This grade of cement achieves a compression strength of 53 N/mm^2 after 28 days.

Out of these three grades of cement, 53 grade is finer than 33-grade and 43-grade cement. Moreover, 53-grade cement is stronger in comparison and can resist sulphate.

1.4.1 Manufacturing Process of Cement

One of the main ingredients in a normal concrete mixture is ordinary Portland cement. However, the production of cement is responsible for approximately 7% of the world's carbon dioxide emissions. Cement production requires raw materials comprising calcareous materials, siliceous materials, and argillaceous materials. Cement can be manufactured using two methods depending upon whether the mixing and grinding of raw materials are done in wet or dry conditions.

1. Wet process
2. Dry process.

1. **The Wet Process** includes the following steps

Raw materials are first collected [calcareous materials (clay and chalk), siliceous materials (siliceous rock), and argillaceous materials (Alumna)] and then mixed in the appropriate proportions. The proportions are dependent on their purity and composition.

The mixture is then burned under the required temperature in a rotary kiln. The temperature in the kiln should be about 1350–1500 °C to break down the materials and allow them to recombine into new compounds.

3–5% of gypsum is added to the cement powder; this addition is done after the clinker cools down to increase the setting time of cement. The cement is then collected and packed in bags. Figure 1.2 shows the flow chart of the wet process of cement.

2. **Dry Process**

In the dry and semi-dry interaction procedure, the required raw materials are squashed in a dry state. At that point, they are prepared in a grinding mill, dried, and grounded to a very fine powder. This dry powder is then mixed and blended in with the compression air. After almost one hour of air circulation, a uniform combination is acquired. At that point, the mixed powder is sieved and applied to measure in a rotating disc. About 12% water by weight is added to allow the mixture to form pellets. The mixture is changed over into pellets by the expansion of water around 12% by weight. Since only minimal equipment is used in the dry process of manufacturing cement, it is economical compared to the wet process (Fig. 1.3).

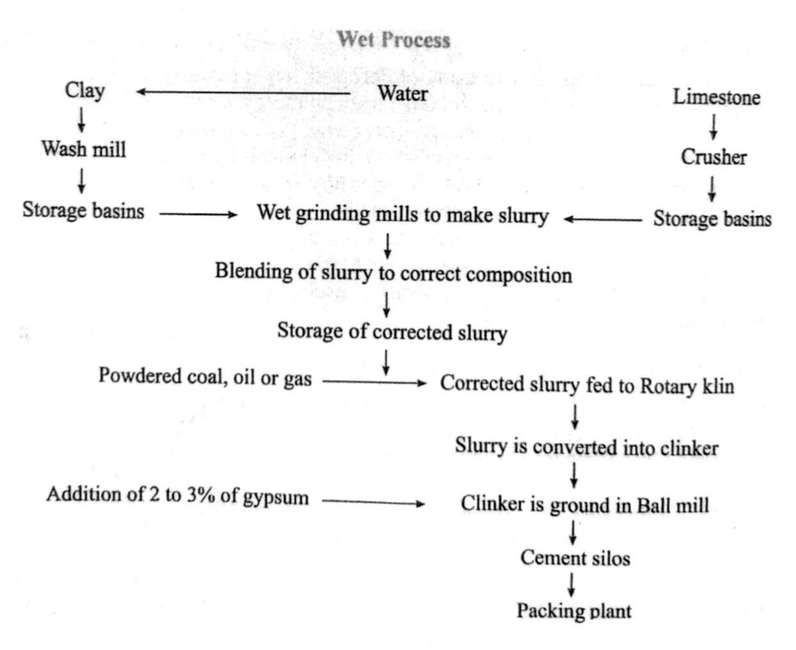

Fig. 1.2 Wet process of manufacturing cement

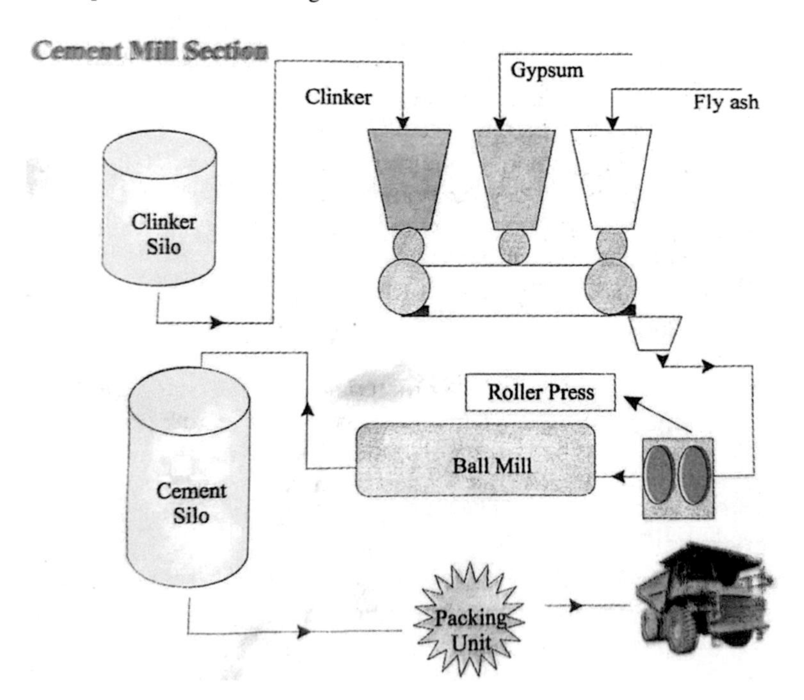

Fig. 1.3 Dry process of manufacturing of cement

Table 1.2 Bogues compounds present in the cement

Name of the Bogues compound	Denoted by
Tricalcium silicate, C3S	C3S
Dicalcium silicate, C2S	C2S
Tricalcium aluminate	C3A
Tetra calcium alumino ferrite	C4AF

1.4.2 Bogues Compounds and Their Significance

Portland cement generally consists of four main compounds, namely tricalcium silicate ($3CaO \cdot SiO_2$), dicalcium silicate ($2CaO \cdot SiO_2$), tricalcium aluminate ($3CaO \cdot Al_2O_3$), and a tetra-**calcium** alumino ferrite ($4CaO \cdot Al_2O_3Fe_2O_3$). The characteristics of these compounds contribute to developing the strength of cement and managing the initial setting time and hardening of the cement (Table 1.2).

The emphasis of C3S in cement plays a vital role in developing the initial strength of cement, while C2S is responsible for developing the strength of cement later during the hydration process of cement. The value of C3A is inversely proportional to the setting time of cement; as the percentage of C3A increases, the setting time of cement decreases. C4AF is responsible for the colour of the cement. Generally, the C3A and C4AF react during the first two to four hours after water is added to Portland cement.

1.5 Tests on Cement

Cement is the key material in developing concrete; thus, it is essential to test the cement before it is used for construction. The tests that need to be done at the construction site are as follows.

1.5.1 Colour Test

The colour of the cement should be uniform and grey. The cement bag should have an ISI symbol with a valid expiry date.

Cement retains its quality for three months after it is packed in bags.

1.5.2 *Lumps Test*

Good cement should not contain any lumps in the cement bag as it affects the hydration process of cement when reacting with water.

1.5.3 *Floating Test*

When some amount of cement is thrown into the water, it should float on the water before sinks into the water. If the cement sinks immediately after immersing in the water, its low quality is revealed.

Chemical tests to check the quality of cement: To determine the quality of cement before use in construction, the following chemical tests are done:

1. Specific gravity of cement
2. Fineness test
3. Consistency test
4. Setting time test
5. Strength test
6. Soundness test
7. The heat of hydration test
8. Tensile strength test
9. Chemical composition test.

Specific gravity of cement: The materials and equipment used to determine the specific gravity of Portland cement are as follows:

i. Portland cement
ii. Kerosene free of water
iii. Standard Le Chatelier flask
iv. Long thistle funnel
v. Thermometer
vi. Balance
vii. Water bath.

The ratio between the weight of a given volume of material and the weight of an equal volume of water is the significance of this experiment. The mass of a unit volume of solids is known as the density of hydraulic cement. It is especially useful when it comes to the design and control of concrete mixtures.

The steps to experiment are as follows:

1. First, the Le Chatelier flask is dried and filled with kerosene oil to a point on the stem between the zero and 1 mL mark.
2. Before any readings are taken, the flask is left at a constant room temperature for a reasonable time.

3. The temperature of the liquid in the flask does not vary more than 0.2 °C.
4. To ensure that the contents of the flask have reached the temperature of the water bath, all readings should be taken after the amount of kerosene in the flask has become constant.
5. The initial kerosene reading in the flask is taken, and the kerosene temperature in the flask is kept constant.
6. The original cement and pan weights are then determined. The flask is steadily filled with a weighted quantity of cement (approximately 65 g) at the same temperature as the oil. It is important to stop splashing and make sure the cement does not stick to the inside of the flask above the kerosene.
7. To speed the introduction of cement into the flask and prevent the cement from sticking to the flask's neck, a vibrating apparatus may be used.
8. When the kerosene level reaches the upper graduations of the flask, more cement should not be added. The flask should be rolled in an inclined position, or gently whirled in a horizontal circle, with the stopper in place.
9. The flask in the water bath is re-immersed, and the final reading is recorded as described in steps 2 and 3. The volume of liquid displaced by the weight of cement (Vc) is used in the test which is the difference between the initial and final readings.
10. The final weights of the cement and the pan weights are noted down. The weight of cement used, Wc, is represented by the difference between the initial and final readings.
11. If the water's specific gravity is equal to one, the cement's basic gravity can be determined as follows: Wc/Vc = Sp. Gr.
12. Kerosene is used to extract the cement and clean the flask. No water should be permitted to enter the flask.
13. The test should be done a minimum of two times for each sample to obtain approximately similar results.

Observations table

Task
Initial flask reading
Initial kerosene temperature
The initial weight of cement sample with pan
Final flask reading
Final kerosene temperature
The final weight of cement sample with pan
Weight of used cement
Volume
The specific gravity of cement

Note The test should be taken for at least three samples and an average value of each specific gravity. The average specific gravity of cement should be around 3.15 for

OPC, while Portland-blast furnace slag and Portland Pozzolana cement can have a specific gravity of around 2.90.

The fineness of the cement test

The following are the effects of cement fineness on concrete:

1. The amount of bleeding in concrete is reduced by increasing the fineness of the cement. This is more evident in concrete that does not have any entrapped air.
2. Lowering the water requirement of concrete by rising the fineness of cement from 2700 to about 4000 cm^2/g. The water requirement of concrete increases as the fineness of cement is increased above a certain stage.
3. Raising the cement fineness increases the workability of non-air-entrained concrete. The effect of cement fineness on workability is much less pronounced in air-entrained concrete.
4. Variations in cement fineness have little impact on the air-void framework in air-entrained concrete or concrete containing no entrained air.
5. With or without entrained air, the 28-day compressive strength of concrete increases as the cement fineness increases.
6. As the fineness of concrete with no entrained air is improved, the static modulus of elasticity decreases slightly after 28 days.
7. The fineness of the cement has an impact on the shrinkage of concrete as it dries. The drying shrinkage increases as the water content increases due to fineness.
8. With rising cement fineness, the resistance of air-entrained concrete to corrosion caused by freezing and thawing decreases. Non-air-entrained concrete exhibits the same pattern, albeit to a lesser extent.
9. In certain building projects, very coarse cement may cause a severe bleeding problem under some circumstances.
10. It is unlikely that changing the fineness of cement that meets current national standards would have a major effect on a field construction issue.

1.6 Aggregate

Aggregates make up nearly 60–75% of the volume of concrete. The bulk of a concrete mixture is made up of fine and coarse aggregates. The most common materials used for this are sand, natural gravel, and crushed stone.

Aggregates are further classified based on shapes as follows:

- Rounded
- Irregular or partly rounded
- Angular
- Flaky.

Aggregates are also divided into two categories as follows:

➢Coarse aggregates
➢Fine aggregates.

1.6.1 Coarse Aggregate

The aggregates which are retained on a 4.75 mm sieve are called coarse aggregates. Generally, the size of coarse aggregate particles is bigger than 4.75 mm.

1.6.2 Fine Aggregate

The aggregates which can pass through a 4.75 mm sieve are called fine aggregates. Generally, the size of fine aggregate particles is smaller than 4.75 mm.

1.7 Water

Hydration is the method of combining water with a cementitious substrate to create a cement paste. More water in the cement paste results in free-flowing concrete with a higher slump; less water in the cement paste results in stronger, more stable concrete. Impurity in the water used to make concrete may cause problems when it sets or cause the structure to collapse prematurely. Water should have a pH of no less than 6. Many different reactions take place during hydration, and they always happen at the same time. The products of the cement hydration process gradually bond together the individual sand and gravel particles, as well as other components of the concrete, to form a solid mass as the reactions progress.

Chapter 2
Admixtures

An admixture is a material added to the cement previously or during its mix to improve its properties. In most developing countries, at least one admixture is being used in the concrete for ideal outcomes. Admixture is the process of adding or removing SCM as a concrete ingredient. Mineral admixture and chemical admixture are the two forms of admixture. To raise or decrease the setting time of concrete, the admixture plays an important role in the strengthening properties of concrete.

Additives, also known as admixtures, have long been used in ordinary Portland cement. Mineral and chemical admixtures are two forms of admixtures that are often blended into Portland clinker or mixed with concrete nowadays. Admixture is the process of adding or removing SCM as a concrete ingredient.

2.1 Mineral Admixture

There are different types of admixtures such as mineral and chemical admixtures that are used in concrete for various purposes. Generally, mineral admixtures such as fly ash, GGBS, hypo sludge, nano-silica, silica fume, marble powder, glass fibres, etc., are used as partial/complete replacement of cement and additives to improve the properties of concrete.

The mineral admixtures are usually used to boost the strength properties of concrete while also reducing the use of natural resources and carbon dioxide emissions into the atmosphere, while chemical admixtures improve the workability and effects on setting time of cement. A few well-known mineral admixtures, such as hypo sludge, ceramic powder, fly ash, GGBS, nano-silica, silica fume, and others, are industrial by-products that could be used as supplementary materials with extreme pozzolana, fine particles, large surface area, and high SiO_2 content, to help enhance concrete strength properties. These are used as an admixture in concrete mixes and have a direct influence on the material's properties. Moreover, the workability of

K. K. Poloju, *Advanced Materials and Sustainability in Civil Engineering*,
SpringerBriefs in Applied Sciences and Technology,
https://doi.org/10.1007/978-981-16-5949-2_2

concrete plays a major role in concrete strength, as it is directly proportional to the grading of concrete. The admixtures are used as an additional component to the components of the concrete. The impact of the admixture uniquely affects the properties of concrete.

2.2 Chemical Admixtures

Chemical admixtures are ingredients that are applied to concrete in the form of powders or liquids to give properties that are not possible with standard concrete mixes. Admixture dosages in regular usage are less than 5% by mass of cement and are applied to the concrete during batching/mixing. The following are the most common forms of admixtures:

1. **Accelerators**: The accelerators are used to increase the rate of hydration (hardening) of concrete.
2. **Retarders**: These additives delay the hydration of concrete and are used in big or difficult pours where partial setting before completion is undesirable.
3. **Plasticizers/superplasticizers (water-reducing admixtures)**: These admixtures improve the workability of concrete, making it easier to place and consolidate. The plasticizers, on the other hand, can be used to minimize the water content of concrete while preserving workability (hence the name "water reducer"). This increases the strength and longevity of the product.
4. **Pigments**: This is used for decorative reasons, and they may be used to alter the appearance of concrete.

Chapter 3
Tests on Hardened Concrete

Concrete should possess both mechanical and durable characteristics. In general, the mechanical and durable properties of concrete are as follows.

3.1 Mechanical Properties of Concrete Are

1. Compressive strength
2. Split tensile strength
3. Flexural strength.

3.1.1 Compressive Strength

The compression testing machine with a capacity of 2000 kN is used to test the cube specimens. The machine's bearing surface must be cleaned, and any loose sand or other material should be removed from the specimen's surface. The specimen is put in the system so that the load is applied to the cubes' opposite sides as a cast, rather than top and bottom. The specimen's axis is carefully centred at the loading frame's base, and then, the load is raised at a steady rate until the specimen's resistance to the rising load broke down and could no longer be maintained. **Generally, the compression strength of concrete cube after curing for 1 day, 3 days, 7 days, 14 days, and 28 days is shown in Table 3.1**.

How to calculate the compressive strength of concrete cube.

At each age, at least three specimens should be examined. If a specimen's strength varies by more than 15% of its average strength, the results of that specimen should be rejected. The compressive strength of concrete cubes is calculated as the average of their specimens.

K. K. Poloju, *Advanced Materials and Sustainability in Civil Engineering*, SpringerBriefs in Applied Sciences and Technology,
https://doi.org/10.1007/978-981-16-5949-2_3

Table 3.1 Percentage of compression strength of concrete cube after curing

S. No.	Compressive strength after days	Percentage of strength gain (N/mm^2) (%)
1	1	16
2	3	40
3	7	66
4	14	90
5	28	99

The cube's dimensions are 150 mm × 150 mm × 150 mm.

The specimen's area (calculated from the specimen's average size) is 22,500 mm^2

The calculation procedure to determine compressive strength of a concrete after 28 days curing is shown below

Assume the applied maximum load = 575 KN or 575,000 N

Compressive strength = (Load in N/Area in mm^2) = 575,000/22,500 N/mm^2

The compressive strength of the concrete cube at 28 days **is 25.6 N/mm²**

Grades of concrete: Generally, the concrete has different grades based on their mix proportion. The characteristic strength of concrete for various grades is shown in Table 3.2.

The characteristic strength of concrete should achieve target mean strength (T.M.S) after 28 days of curing. The below formula is used to determine the Target Mean Strength (T.M.S) = [Fck + (K * S)].

Where fck = characteristic strength of concrete,

K = coefficient ($K = 1.65$),

S = standard deviation ($S = 4$ for M20 & M25 and $S = 5$ for M30 and more).

Table 3.2 Grade and characteristic strength of concrete

S. No.	Grade of concrete	Characteristic strength after 28 days,
1	5	5
2	7.5	7.5
3	10	10
4	15	15
5	20	20
6	25	25
7	30	30
8	35	35
9	40	40
10	50	50

3.1.2 Split Tensile Strength

The cube specimens are placed through their paces on a 200 tonne compression testing unit. In the case of cylindrical specimens, the test is carried out by positioning the specimen horizontally between the loading surfaces of the compression testing system for split tensile strength, and the specimen's axis is carefully aligned at the centre of the loading frame. The load is raised at a steady rate until the specimen's resistance to the rising load broke down, and it could no longer be maintained. The maximum load that could be applied to the specimen is registered.

The split tensile strength is obtained by

$$2P/(\pi * LD)$$

where

P is the load on the cylinder,
L is the length of the cylinder,
D is the diameter.

3.1.3 Flexural Strength

The modulus of rupture represents the specimen's flexural strength. One point loading is the approach used in the testing. The test specimen should be turned on its side and centred on the bearing blades concerning its section moulded. At one point on the supports, the blades on which the load has been applied must make contact with the upper surface. The severe fibre stresses on the tensile at the point of failure are what give bearings their strength.

If "*a*" equals the distance between the line of fracture and the closer support calculated on the central line of the tensile side of the specimen in cm, the following formula is used to determine "a" to the nearest 0.05 M pa.

$$F = p \times l/b \times d^2$$

The above formula is used when the value of "*a*" is greater than 20 cm for 150 mm size specimen or 13.3 cm for 100 mm specimen or

$$F = 3P \times a/b \times d^2$$

The above formula is used when the value of "*a*" varies between 17 cm to 20 cm for 150 mm size specimen or 11 cm to 13.3 cm for 100 mm specimen.

Where

$B =$ the specimen's estimated width in cm.
$D =$ the specimen's measured depth in centimetres at the point of failure.

L is the length of the span on which the specimen is supported in centimetres.
M is the maximum load in kilograms that can be added to the specimen.

The compressive strength, split tensile strength, and flexural strength of the cubes, cylinder, and prism are determined by dividing the load at failure by the cross-sectional area of the cubes, cylinder, and prism.

3.2 Durability Properties of Concrete

1. Acid attack
   I. Preparation of H_2SO_4 solution
   II. Preparation of HCL solution
2. Sorptivity

3.2.1 Acid Attack

The chemical resistance of concrete is investigated by immersing it in an acid solution and exposing it to chemical attack. After a 28-day curing cycle, each batch's specimen is extracted from the curing tank and cleaned with a soft nylon brush to extract poor reaction products and loose materials. The initial mass, as well as the value of the body diagonal measurements, is to be measured. Every batch of concrete with three specimens is immersed in a solution of 5% H_2SO_4 and 5% HCl. For 3, 7, and 28 days, the weight loss, compressive strength loss, and acid durability factors are checked.

Preparation of 5% H_2SO_4: (An example is shown on how to prepare solution)
Density of H_2SO_4 = 1.83 g/cc, volume of H2SO4 taken = 100 mL
H2SO4 (98% purity) mass = 100 × 1.835 = 183.5 g
H2SO4 actual mass = 98/100 × 183.5 = 179.83 g
Actual mass H_2SO_4/(mass H_2SO_4 + mass water) = 5% H_2SO_4
X = 3413 mL of water, 5/100 = 179.83/(183.5 + x)

That is, to make a 5% H_2SO_4 solution, 3413 mL of water is applied to 100 mL of 98% H_2SO_4, as shown in Fig. 3.1.

3.2.2 Preparation of 5% HCL: (An Example Shown on How to Prepare Solution)

5% HCl preparation: HCl volume = 100 mL
100 × 1.18 = 118 g of HCl (36.5% purity).
H_2SO_4 mass is 36.5/100 × 118 = 42.07 g.
Actual mass HCl/(mass HCl + mass water) = 5% HCl.

Fig. 3.1 Immersion of cubes in 5% H_2SO_4

Water: $5/100 = 42.07/(118 + x)\ X = 731$ mL.

To make a 5% HCl solution, 731 mL of water is added to 100 mL of 36.5% HCl, as shown in Fig. 3.2.

After 3, 7, and 28 days of immersion, the mass and compressive strength are to be assessed. Throughout the reliability test, the solutions' normalcy is preserved. After every 10 days, the normality of the solution is checked; if there is a drop in normality, the amount of acid intake is calculated and substituted with that amount. Each specimen is removed from the bath, cleaned with a soft nylon brush, and rinsed in tap water before being tested. This procedure cleans the specimens by removing loose surface material.

The durability factors suggested by the ASTM philosophy can be used to determine the resilience of concrete specimens to hostile environments like acid attacks

Fig. 3.2 Immersion of cubes in 5% HCL

(666–1997). The "acid durability factors" are calculated in terms of relative strength in the current study. The relative strengths are always measured against the value from the previous 28 days (i.e. at the start of the test).

The "acid durability factors" (ADF) are determined in the following manner.

$$\mathrm{Sr} = N/M(\mathrm{ADF})$$

where Sr denotes relative strength at *N* days, and

N is the number of days that the durability factor is needed.

M—The number of days before the exposure is to be halted. In this case, *M* is 28.

3.2.3 Sorptivity Study

After drying the specimens in an oven at a temperature of 100(±), 10 °C, the sorptivity test is to be performed on all specimens with dimensions of 15 × 15 × 15 mm. The samples are to be held in the freezer until the weight loss is minimal. Water impermeability of the lateral faces of the samples is also included in their preparation, minimizing the effect of water evaporation. The test began with the weight of the samples being registered, and then, they put in a recipient with enough water to submerge them at around 5 mm. The samples are taken from the recipient after a predetermined period to begin weight registration. The samples' surface water is removed with a wet cloth until it is weighed. The samples are immediately substituted in the recipient after weighing until the next measuring period. The procedure is repeated at different intervals, including 15 min, 30 min, 45 min, 1 h, 24 h, 3 days, 7 days, and 28 days.

The relationship between cumulative water absorption (kg/m^2) and the square root of exposure time (t0.5) deviates from linearity during the first few minutes due to low initial surface tension and buoyancy effects. Thus, only the portion of the curves for the exposure time of 15 min to 72 h, where the curves are continuously linear, is used to calculate the sorptivity coefficient.

The following expression is used to calculate the sorptivity coefficient (k):

$$\frac{W}{A} = k\sqrt{t}$$

where W is the volume of water adsorbed in (kg),

A is the cross section of the specimen in contact with water (m^2),

t is the time (min), and

k is the specimen's sorptivity coefficient (kg/m^2/min0.5).

Chapter 4
Industrial Wastes as a Cement Substitute in Concrete Production

4.1 How Global Warming Relates to the Manufacturing Process of Cement

Concrete is a necessary component in all types of construction projects. Cement plays a vital role in the construction of concrete. However, the demand for cement in developing countries is rapidly growing these days. The processing of OPC necessitates a great deal of energy and has a large carbon footprint, accounting for about 7% of global CO_2 emissions per year. This contributes to the greenhouse effect, which is a serious problem in every part of the world. To create environmentally friendly concrete, it is important to use alternative and renewable materials in the production of cement. Furthermore, a variety of industrial wastes can be used to replace cement and aggregates. The profitable use of these industry by-products decreases landfilling and emissions. Furthermore, the advantages of using various wastes produced by factories as supplementary cementitious materials to the concrete ingredients are to determine the optimum percentage of use in concrete. With the partial replacement of cement and aggregates, the results obtained from various tests using various waste products indicate that these have better properties to achieve good strength.

Although developed, industrialized countries are being urged to reduce environmental pollution and their share of global resource use, the developing countries must avoid the mistakes of the past. This problem is exacerbated by the fact that cement production and fly ash generation in China and India are expected to rise dramatically in the coming decades. Concrete is in high demand around the world, with demand expected to double in the next 30 years. We can meet this demand without raising greenhouse gas emissions by using supplementary cementitious materials to substitute as much cement as possible in concrete. We can thus save energy and resources, minimize CO_2 emissions, and reduce the negative environmental effects.

K. K. Poloju, *Advanced Materials and Sustainability in Civil Engineering*,
SpringerBriefs in Applied Sciences and Technology,
https://doi.org/10.1007/978-981-16-5949-2_4

Table 4.1 Chemical composition of cement and different industry wastes

Chemical compound	Cement %	Hypo sludge	Ceramic powder %	Fly ash %	GGBS %
Calcium oxide	60–67	46.2	4.47	4.00	32.6
Silicon dioxide	17–25	09	63.28	60.11	34.06
Aluminium oxide	3.0–8.0	3.6	18.28	26.53	20
Iron oxide	0.5–6.0		4.35	4.25	0.8
Magnesium oxide	0.1–4.0	3.33	0.75	1.25	7.89
Potassium and sodium oxide	0.4–1.3	4		0.22	
SO_3	1.3–3.0			0.35	0.9

4.2 Various Supplementary Cementitious Materials Used in Concrete

The most widely used supplementary cementitious materials are as follows:

1. Hypo sludge
2. Fly ash
3. GGBS
4. Ceramic powder
5. Marble powder
6. Seashells.

The above-mentioned waste products would otherwise end up in landfills contributing to the environmental value. Energy is critical to the development of developing countries. The importance of using industrial waste is very significant in the sense of the low availability of non-renewable energy resources combined with the requirements for large quantities of energy for construction materials like cement. The chemical composition of cement and different industry wastes is shown in Table 4.1.

Chapter 5
Hypo Sludge as Partial Replacement of Cement

5.1 Introduction

Various types of waste are generated by the paper industry when processing paper. Hypo sludge is the term for the preliminary waste from the paper industry which contains low calcium content, high calcium chloride content, and low silica content. Because of its silica and magnesium content, hypo sludge behaves like cement. Silica helps in developing the bonding between aggregates in concrete. Hypo sludge is used as a substitute cementitious material for partial replacement in high-performance concrete. The strength of the concrete will increase when this waste is used to mix concrete, and another advantage is that the cost of cement production will be reduced. Raw hypo sludge generated from a paper factory is shown in Fig. 5.1.

5.2 Formation of Hypo Sludge Concrete

Hypo sludge concrete is made up of similar materials that are used to develop conventional concrete. However, the hypo sludge is used as a partial replacement for cement during the mixing process. The coarse and fine aggregates make up about 75–80% of the mass, equivalent to regular concrete. The aggregates are mixed with a binder material (cement + hypo sludge) along with the required quantity of water. Chemical admixtures such as superplasticizers may be applied during the mix to change their properties to suit the project's requirements. Superplasticizers minimize the amount of water in the mixture, making it much more workable. The chemical composition of hypo sludge and comparison with cement are shown in Tables 5.1 and 5.2.

K. K. Poloju, *Advanced Materials and Sustainability in Civil Engineering*,
SpringerBriefs in Applied Sciences and Technology,
https://doi.org/10.1007/978-981-16-5949-2_5

Fig. 5.1 Raw hypo sludge disposal

Table 5.1 Properties of raw hypo sludge

S. No.	Constituents	% Present in hypo sludge
1	Moisture	56.8
2	Magnesium oxide (MgO)	03.3
3	Calcium oxide (CaO)	46.2
4	Acid insoluble	11.1
5	Silica (SiO_2)	09.0

Table 5.2 Comparison between cement and hypo sludge

S. No.	Constituents	Cement (in %)	Hypo Sludge (in %)
1	Lime (CaO)	62	46.2
2	Silica (SiO_2)	22	0.9
3	Alumina	05	3.6
4	Magnesium	01	3.33
5	Calcium sulphate	04	4.05

5.3 Chemical Composition of Hypo Sludge

5.4 Solid Waste Produced by the Paper Industry

Table 5.3 shows the solid wastes produced while manufacturing paper, which are divided into two categories: combustible (organic) and non-combustible (inorganic). Table 5.4 shows the amount of waste generated.

Table 5.3 Major sources of solid waste generated

Sources	Type	Wastes
Raw material preparation, paper machine, ETP	Combustible wastes (organic)	Straw and agro-waste dust, pith, fibrous sludge, and primary and secondary sludge
Bleach liquor preparation plant and steam boiler	Non-combustible wastes (inorganic)	Hypo sludge, fly ash, and cinder

Table 5.4 Quantity of solid waste generated

S. No.	Solid waste	Quantity, kg/t Paper
1	Raw material preparation rejects (bagasse pith, straw dust, etc.)	45–480
2	Process rejects	20–100
3	Hypo sludge*	35–65
4	ETP sludge	35–180
5	Cinder/fly ash	50–330

5.5 The Significance of Using Hypo Sludge in Concrete

Natural materials, on average, require a higher level of quality control than recycled materials. However, as natural resources become scarcer and the costs of disposing of building waste and other waste materials rise, recycling will become more economically viable. Paper processing creates a huge amount of solid waste which is formed by the separation of torn, low-quality paper fibres from waste sludge. Every year, this paper mill sludge uses a considerable amount of local landfill space and leads to severe air pollution issues. It is important to decontaminate these industrial wastes to reduce disposal and contamination issues. The use of hypo sludge instead of Portland cement decreases greenhouse gas emissions in concrete, as one tonne of Portland cement contains approximately one tonne of CO_2 as discussed earlier. Since global Portland cement production is projected to hit nearly 2.1 billion tonnes by 2020, replacing the cement with 30% of hypo sludge will significantly reduce global carbon emissions.

Illustration-1—Advancement of Physical Properties of Hypo Sludge Concrete for M 20 Grade of Concrete

Kiran et al. [1] investigated the M20 grade of concrete utilizing hypo sludge as a partial substitute material of cement. Different mould sizes are used to determine the strength properties of concrete such as compressive strength, split tensile strength, and flexural strength using cube moulds (150 × 150 × 150 mm), cylindrical moulds (150 × 300 mm), and prism moulds (150 × 150 × 700 mm). These specimens are cured underwater for 7 and 28 days. Similarly, the acid tests are carried out using the cube moulds after immersing in 5% HCL and 5% H_2SO_4 for assessing the durability properties. Additionally, the sorptivity test is also conducted. The test outcomes show

Table 5.5 Details of specimens cast

S. No.	Grade of concrete	Type of concrete	% of hypo sludge	No. of cubes cast	No. of cylinders cast	No. of prisms cast
				150 × 150 × 150 mm	150 × 300 mm	150 × 500 mm
1	M20	Without hypo sludge	–	15	3	3
		With hypo sludge	10%	15	3	3
			20%	15	3	3
			30%	15	3	3
			40%	6		
			50%	6		
Total				72	12	12

that the utilization of hypo sludge in concrete has improved the strength properties compared to conventional concrete.

In their study, the experiment involved casting and measuring 72 cubes, 12 cylinders, and 12 prisms to determine mechanical properties such as compressive strength, split tensile strength, and flexural strength for M20 grade concrete at various cement replacements. Table 5.5 shows the specimen's specifics. For durability tests, a total of 24 cubes are cast for M20 grade concrete, with 12 cubes for H2SO4 testing and 12 cubes for HCL testing. Cubes are also used in sorptivity experiments.

Materials that are used in this investigation

Cement—The cement used in this project is a 53-grade cement that complies with IS:12269. The cement used has a specific gravity of 3.15. Throughout the experimental program, the same constituent is used to make the cement. The initial setting time is 30 min.

Fine Aggregate—In this experimental scheme, river sand conforming to IS 383 Zone 2 sand is used. The fine aggregate had a specific gravity of 2.62.

Coarse Aggregate—This experimental software used machine crushed granite that complied with IS 383 and IS 2386. The coarse aggregate had a specific gravity of 2.82.

Water—Potable water is used in this investigation.

Hypo sludge—Hypo sludge is the term for the preliminary waste generated from the paper industry. Hypo sludge is used as a substitute cementitious material for partial replacement in high-performance concrete. The strength of the concrete will increase, and the cost of the concrete will reduce as a result of the use of this waste.

Table 5.6 Mix proportion of hypo sludge concrete

Water (kg/m^3)	Cement (kg/m^3)	Fine aggregate (kg/m^3)	Coarse aggregate (kg/m^3)
125	250	711	1420

Admixture (Conplast SP430)—Since there is a reduction in water content when we substitute cement with hypo sludge, we used 0.5% of the cement weight to increase workability in this analysis (Table 5.6).

Curing procedure of concrete specimens

After the casting is completed, all specimens are kept in the same environment for 24 h, with a temperature of 27 ± 2 °C and relative humidity of 90%. The specimens are taken out of the mould and immersed in clean, freshwater until they are ready to be examined. The water in which the cubes are submerged is held at a constant temperature of 27 ± 2 °C. The specimens are cured for seven and twenty-eight days, respectively.

Main findings: Figures 5.2, 5.3, and 5.4 show the mechanical properties of conventional concrete and hypo sludge concrete (Fig. 5.5).

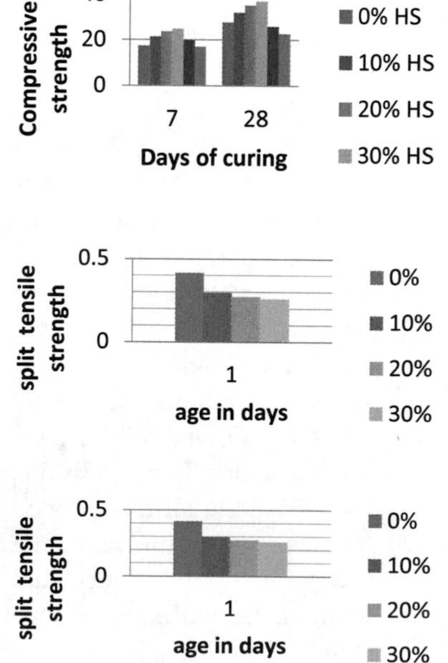

Fig. 5.2 Compression strength of conventional concrete and hypo sludge concrete

Fig. 5.3 Split tensile strength of conventional concrete and hypo sludge concrete

Fig. 5.4 Flexural strength of conventional concrete and hypo sludge concrete

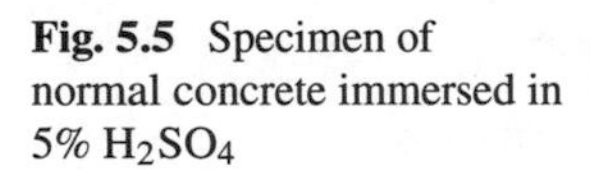

Fig. 5.5 Specimen of normal concrete immersed in 5% H_2SO_4

5.6 Conclusions

The current study examined the impact of hypo sludge on the strength and durability of normal concrete. The following conclusions are drawn as follows:

1. The substitution of cement with hypo sludge ranges from 0 to 50%. When 30% of the cement is substituted with hypo sludge, the optimum compressive strength is reached in 28 days. For a 30% substitution of standard concrete, the percentage improvement in compressive strength is 33.69.
2. Flexural strength is improved from 0 to 30% replacement, with a percentage increase of 13.09 for 30% replacement relative to regular concrete.
3. After 28 days, the percentage mass loss for hypo sludge with 10% replacement is 1.705% in 5% sulphuric acid (H2SO4), which is less compared to normal concrete. After 28 days, the percentage mass loss for hypo sludge with 10% replacement is 1.1% in 5% hydrochloric acid (HCl) which is less when compared to other replacements.
4. As opposed to hypo sludge concretes, the percentage loss of compressive strength for cubes immersed in 5% H2SO4 and 5% HCL after 28 days is less for regular concrete.
5. The acid durability factor (ADF) for normal concrete is higher for cubes held in 5% H_2SO_4 and 5% HCl, with values of 88.36 and 89.95, respectively. As a result, ordinary concrete is long-lasting.
6. As opposed to all other concretes, the coefficient of sorptivity for 30% hypo sludge concrete is lower (0, 10, and 20%).
7. It can thus be concluded that hypo sludge concrete meets the compressive strength, flexural strength, and sorptivity criteria. However, it has struggled to meet the criteria for long-term durability.

Reference

1. K.P. Kiran et al., Advancement in physical properties of hypo sludge concrete. Int. J. Sci. Eng. Res. (IJSER) **7**(9) (2016). ISSN 2229-5518 with Paper ID: I090391. https://www.ijser.org/researchpaper/Advancement-in-Physical-Properties-of-HypoSludge-Concrete.pdf

Chapter 6
Marble Powder as Partial Replacement of Cement

6.1 Introduction

Marble is a metamorphic rock, which means it has changed over time. Marble has been used in the building, decoration, and decoration of houses and palaces for a long time. When limestone is subjected to intense pressure and heat, marble is made. Metal calcite, as well as other minerals, makes up the majority of marble (e.g. quartz, graphite, mica, iron oxides, clay minerals, and pyrite). Any concrete portion can be partially replaced with marble waste (cement).

Human health and the atmosphere are also harmed by industrial waste (marble powder waste). When marble quarries cut stones, marble powder waste is produced. Marble powder waste is deposited underground or buried in landfills, ponds, or landfills. Marble powder waste pollutes the air, water, and soil. As a consequence, a solution for the disposal and utilization of these wastes is needed.

A few researchers have been working with marble powder as a partial substitute for cement, but there is a dearth of investigations on the strength and durability of this alternative. The mechanical and long-term properties of specimens with and without marble powder are investigated using the properties of concrete. Nowadays, various industrial wastes are being used in research areas that include a variety of fields, including civil engineering and building materials.

1.1 Ingredients present in concrete
1.2 Chemical composition of marble powder compared to cement
1.3 Advantages of using marble powder as a partial replacement as cement in concrete.

K. K. Poloju, *Advanced Materials and Sustainability in Civil Engineering*,
SpringerBriefs in Applied Sciences and Technology,
https://doi.org/10.1007/978-981-16-5949-2_6

6.2 Illustration-2—Determination of Strength Properties of Concrete Using Marble Powder as a Partial Replacement as Cement in Concrete

As opposed to traditional concrete, marble powder has a large effect on its properties because it has the most intense and workability. The benefits of using marble powder in concrete technology are discussed in this article. The primary aim of this project is to assess the feasibility of using marble powder as a concrete substitute to improve quality properties. The marble dust powder increased the M25 concrete grade at 0%, 10%, 20%, and 30% by the weight of concrete. Slump test, compression test, split tensile test, flexural tests, and rebound hammer tests are used to evaluate the properties of fresh concrete and equate them to standard cement. This investigation used a variety of mould sizes to determine the quality properties of cement. Both experiments are carried out after 7 and 28 days of water curing to establish comparative thinking on the strength properties of concrete. The results of the tests show that the display of pressure and split tensile strength, as well as resilience, has improved and that a 20% replacement is optimal.

6.3 Materials Used

Ordinary Portland cement, aggregates, water, and marble powder are among the materials used in the investigation. To assess the workability and strength properties of concrete, various equipment and moulds are used. The experiment uses marble powder as a cement substitute at different replacement levels varying from 0 to 30% with a 10% interval for M25 grade. The ratio of the ingredients is 1:1:2, with a W/C of 0.5.

6.4 Methodology

Standard moulds of 150 mm × 150 mm × 150 mm, 150 × 300 mm, and 100 × 100 × 500 mm are used to assess compressive strength, split tensile strength, and flexural strength, respectively. These standard moulds are mounted in such a way that there are no gaps between the mould plates. If there are some minor holes, plaster of Paris is used to fill them. After that, the moulds are oiled and stored in preparation for casting. Similarly, after casting all specimens, cubes are cast to assess a few toughness properties. All specimens are demoulded and moved to a curing tank, where they are submerged in water for the duration of the curing process.

Casting and testing of M25 grade concrete with and without marble powder are parts of the experiment undertaken to check the characteristics of concrete. This provides a thorough description of the concrete design for M25 grade and marble

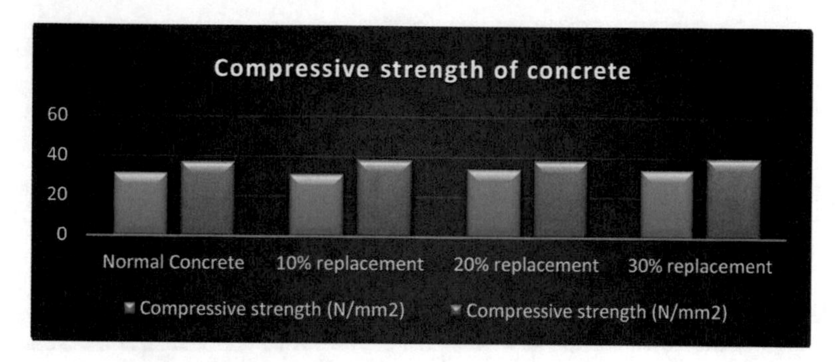

Fig. 6.1 Compression strength of conventional concrete and marble powder concrete

powder concrete. The appropriate ingredients are mixed in a mixing machine with a sufficient mix proportion for 3–5 min during the mix. Fresh concrete is poured into cube moulds with a smooth surface finish. After 24 h, the moulds are demoulded and stored for curing in various regimes, with tests conducted after 7, 28, and 90 days to determine the necessary characteristics.

6.5 Significance of Using Marble Powder in Concrete

Most developing countries strive to reduce pollution of the environment and to make more efficient use of the world's resources, such as natural resources and energy. Cement is gaining popularity around the world, with some reports estimating that demand will double in the next 30 years. Therefore, supplementary cementitious materials are used in concrete to solve this issue. The processing of marble powder produces a lot of solid waste. It means that the broken, low-quality bits are removed and discarded. This broken marble powder waste consumes a significant amount of local landfill space per year and contributes to genuine land pollution issues. Therefore, it is required to attempt various research works to describe the technological and environmental advantages of using supplementary cementitious materials, as well as the drawbacks, applications, and specifications simply and concisely.

Main findings: Figures 6.1, 6.2, and 6.3 show the mechanical properties of conventional concrete and marble powder concrete.

6.6 Conclusion

1. The results indicate that marble powder could replace cement and have more quality compared to normal concrete, increasing compressive strength by 30%.
2. For concrete, its tensile strength increases up to 20% substitution, and for 30% substitution, there can be a slight reduction in quality, so 20% seems ideal.

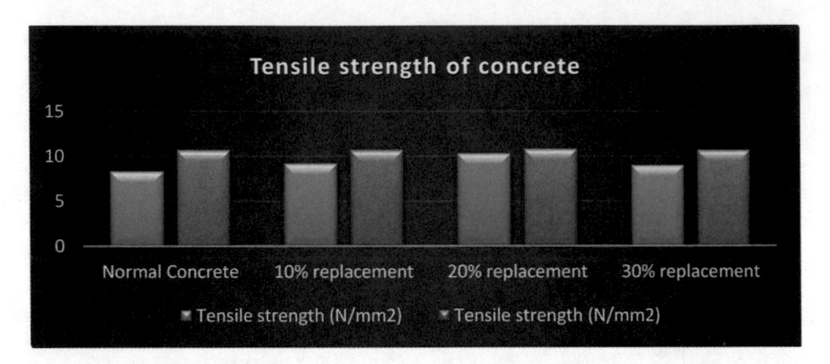

Fig. 6.2 Tensile strength of conventional concrete and marble powder concrete

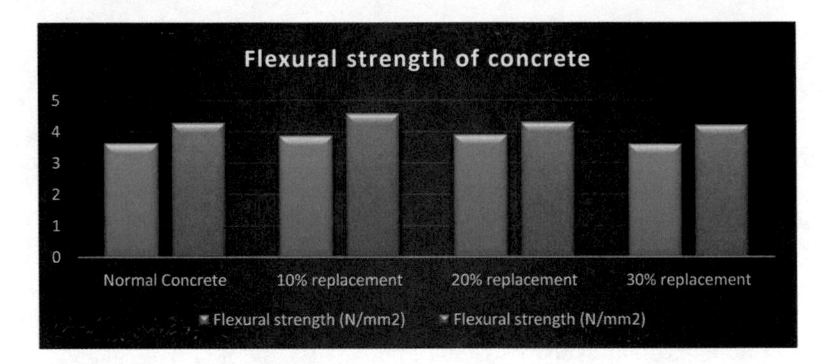

Fig. 6.3 Flexural strength of conventional concrete and marble powder concrete

3. As far as Flexural strength is concerned, 10% substitution yields the best results both on 7 days and 28 days, likewise 20% substitution yields a better quality than ordinary cement, however at 30% substitution the quality decreases more than typical cement, with 20% said to be the best.
4. It could be said that marble powder can be used as a partial substitute for cement with up to 20% of concrete being substituted.
5. This enables the construction industry to make feasible advancements.

Chapter 7
Ceramic Powder as Partial Replacement of Cement

7.1 Introduction

Ceramic tiles are now being used in several industries for different forms of building. Approximately, one hundred million tons of ceramic tiles are manufactured each year by various factories around the world. However, there are still waste materials in the overall ceramic tile production, ranging from 20 to 30% of the total ceramic tile production [1]. As a result, the advancement in concrete manufacturing has a positive effect on the environment by reducing natural resource use [2]. Furthermore, there are several benefits of using ceramic waste powder as a partial substitute for ordinary Portland cement (OPC) in concrete mixing.

7.2 Significance of Using Ceramic Powder in Concrete

Ceramic is produced by a variety of industries around the world. However, since there is no need for recycling ceramic waste, it is sadly ending up in a landfill. As a result, ceramic waste has a detrimental influence on the environment and human health, causing severe diseases. As a result, using ceramic waste powder in concrete will be a perfect way to protect the atmosphere while still having other benefits. Natural raw materials are used extensively in the concrete industry. This has implications for the environment, energy use, and land, as well as for urban growth and the destruction of mines and old structures.

K. K. Poloju, *Advanced Materials and Sustainability in Civil Engineering*,
SpringerBriefs in Applied Sciences and Technology,
https://doi.org/10.1007/978-981-16-5949-2_7

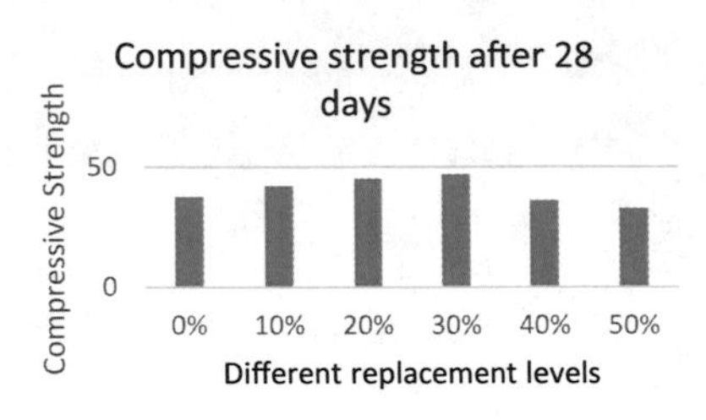

Fig. 7.1 Compression strength of conventional concrete and ceramic powder concrete

7.3 Illustration-3—Advancement of Ceramic Powder in Concrete

The advantages of ceramic waste in the construction industry are discussed by Al-Ruqaishi et al. [3] who justify and verify the fresh properties of concrete using tests such as the slump test, the basic gravity of cement, fine aggregate, and coarse aggregate, as well as water absorption of aggregates. The cubes and cylinders with dimensions of 150 × 150 × 150 mm and 150 × 500 mm are cast with and without 10%, 20%, and 30% ceramic powder, respectively. Axial compression and tensile tests of cubes and cylinders are performed for 7 days and 28 days of curing in water to get a general understanding of the mechanical properties of concrete. To determine the durability of ceramic-based concrete, a sorptivity test is performed, and it is found that the use of ceramic powder in concrete increased the compression strength and durability of the material.

7.4 Materials That Are Used

The materials used in the investigation included ordinary Portland cement, aggregates, water, and ceramic powder. Slump cone and different sizes of moulds are used to determine the workability and strength properties of concrete. For M25 grade, the experiment used ceramic powder as a cement substitute at various substitution levels ranging from 0 to 30% with a 10% interval. The water–cement ratio (W/C) is 0.45, and the mix proportion is 1:0.75:1.5. To assess the strength properties of concrete, cubes, cylinders, and prisms are cast and water cured for 7 and 28 days, respectively.

Main findings: Figures 7.1, 7.2, and 7.3 show the mechanical properties of conventional concrete and ceramic powder-based concrete.

7.5 Conclusion

Effect of ceramic powder on strength properties: In their research, M30 is produced with various levels of ceramic powder replacement as cement.

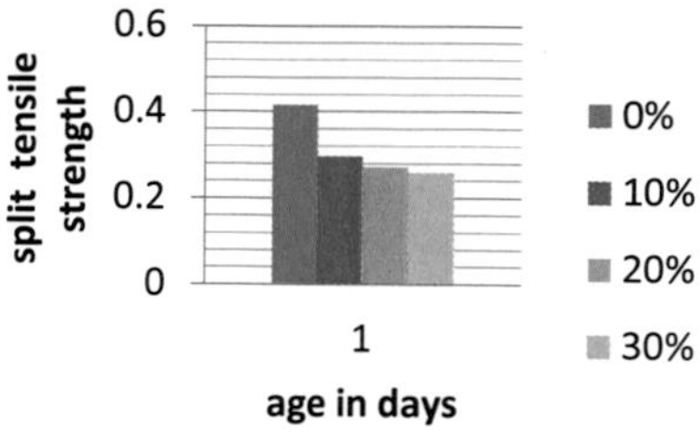

Fig. 7.2 Split tensile strength of conventional concrete and ceramic powder concrete

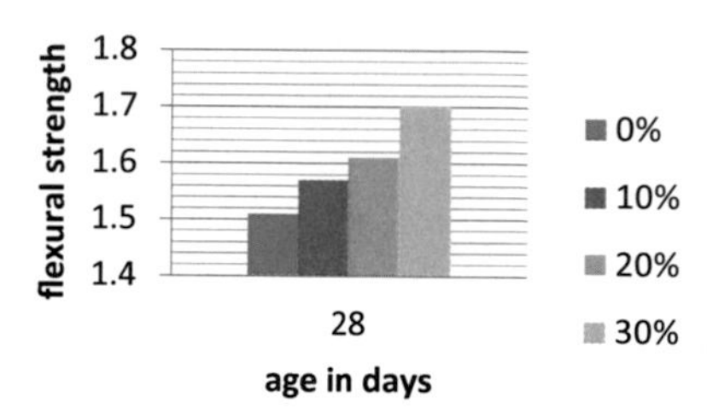

Fig. 7.3 Flexural strength of conventional concrete and ceramic powder concrete

1. The compressive strength increased significantly after 28 days with the addition of 10%, 20%, and 30% ceramic powder, respectively, and then began to decrease to 40 and 50%. As a result, it is concluded that a 30% addition of ceramic powder is optimal.
2. Adding ceramic powder to concrete increases the flexural strength by up to 30% of the ceramic powder blend.
3. As the amount of ceramic powder in the mix increases, the split tensile strength decreases. The tensile strength of 20% ceramic powder is strong, according to the results.

Effect of ceramic powder on sorptivity: Initially, all of the batches showed a slight increase in water absorption, but after 14 days, there is a decrease.

Just one or two specimens out of every three in each batch displayed a small increase in weight.

While there is a lot of contrast with the other ceramic powder percentages, 20% ceramic powder has less sorptivity than the others. It could be concluded that the use of ceramic waste powder in concrete as a partial substitute for cement has a variety of benefits, including lower concrete costs and less landfill waste.

References

1. Ceramic World (2017) World production and consumption of ceramic tiles [online]. Available: https://www.ceramicworldweb.it/cww-en/statistics-and-markets/world-production-and-consumption-of-ceramic-tiles-4/ [2 January 2019]
2. CE, CR (2018) Development in concrete technology [online]. Available: https://www.concreteshowindia.com/blog/development-in-concrete-technology/ [2 January 2019]
3. Al-Ruqaishi AZM, Allamki MSHA, Poloju KK (2019). The advancement of ceramic waste in concrete. Int. J. Advances Appl. Sci. **6**(11), 102-108

Chapter 8
Seashell as Partial Replacement of Cement

8.1 Introduction

Seashells have tough protective exoskeletons. These exoskeletons are mostly composed of calcium carbonate and a small amount of protein, making them common in seashores and valleys. Generally, seashells are regarded as redundant and cannot be used as key ingredients in manufacturing. Due to their chemical composition, seashells can be used in place of the usual fine aggregates like sand and crushed stones, but they also contain calcium oxide, silicon oxide, and alumina, which are all big elements in fine aggregates and cement as well.

8.2 Illustration-4—Advancement of Physical Properties of Concrete Using Seashell as a Partial Replacement of Fine Aggregate in Concrete

The possibility of replacing seashells with aggregate was rarely investigated. Thus, a detailed experiment has been carried out by Kiran et al. to determine the mechanical properties of concrete as a partial replacement of fine aggregate with seashell in different replacement levels. Based on their studies, they demonstrate how to achieve better results when using seashells as a partial fine aggregate that includes workability characteristics. By incorporating seashells into construction, local concrete makers will be able to realize the potential benefits of using seashells in local construction.

K. K. Poloju, *Advanced Materials and Sustainability in Civil Engineering*,
SpringerBriefs in Applied Sciences and Technology,
https://doi.org/10.1007/978-981-16-5949-2_8

8.3 Ingredients Used in Seashell Concrete

To carry out this study, different materials were used.

1. **Ordinary Portland Cement (OPC)**: 53-grade cement is used, and basic tests are carried out such as setting time and fineness.
2. **Fine Aggregate**: The experimental programme used Zone 2 sand as fine aggregate. By using a 4.75 sieve, the fine aggregate is sieved until it reaches the desired weight after being collected from the local source.
3. **Coarse Aggregate**: A machine-crushed granite sample was used in an experimental programme, sieved with a 20 mm and a 14 mm sieve, and analysed.
4. **Water**: Water used in concrete mixing should be pure, which means it should not smell or have any colour to it. The water mixes the concrete properly and binds its components. The quality of water influences many characteristics, such as workability, concrete's strength, permeability, and durability.
5. **Admixture (Conplast SP430)**: An admixture of 0.5% of cement weight is used to improve workability.
6. **Seashells**: It is easily found in any marine or coastal area. Seashells are very versatile and can be used for a variety of purposes, such as decoration, jewellery, or even as an admixture in concrete. It was burned to eliminate organic compounds and crushed before being added to the concrete mix. Similar to the sieves used to sieve the fine aggregate, the seashell was sieved using a 4.75 mm sieve.

Methodology: During this experiment, seashells were used to calculate the impact they have on concrete properties. The emphasis in this work is on compaction strength, which is linked to other properties such as durability. A rotary mixer is used to mix all the required materials with water for 3–5 min. 0%, 10%, 20%, and 30% of seashells were replaced in the cubes to find out the strength properties. The moulds are filled with freshly mixed concrete. The top surface should be well finished, as should the size of the specimens, such as cube moulds (150 mm × 150 mm × 150 mm). Demoulded specimens should be kept in water for seven and fourteen days after they have been cast.

Main Findings and Discussion: According to the obtained results, replacing the fine aggregate with seashells resulted in little impact on the compressive strength of concrete. It has been demonstrated in our research and that of previous studies that concrete becomes more compressive with the replacement of seashells, but usually only in specific amounts. Three different percentages of partially replaced seashells were observed in their experiments, 10, 20, and 30% at two different curing times, 7 and 14 days, respectively, compared to conventional concrete (Fig. 8.1).

The compressive strength of concrete cubes with 0% replacement showed similar properties to other replacements after 7 days and 14 days of curing, i.e. 43.37 and 50.6, respectively. The average strength after 10% replacement is 44.75 days for 7 days and 50.43 days for 14 days. Likewise, the strength obtained for 20% and 30% replacements is 44.6 and 50.6, 47.9 and 47.7 14 for 7 and days, respectively. In

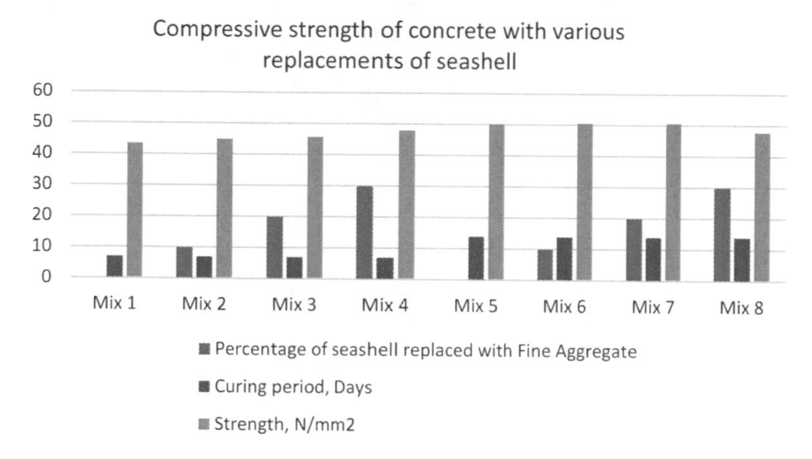

Fig. 8.1 Compressive strength of concrete with various replacements of seashell

comparison with other replacements, a decrease in strength for 14 days was observed. Thus, 20% replacements might be better to achieve better strength properties.

8.4 Conclusion

For conventional concrete, the compressive strength is 50.06 MPa after 14 days. Meanwhile, the compressive strength with 20% replacement is 50.62 MPa, and the replacement with 10% is 50.43, which is higher than conventional concrete 50.06, but lower than strength with 20% replacement.

It can be concluded from a sustainability perspective that partial replacement of seashells with fine aggregate is a viable alternative to conventional fine aggregate due to its increased compressive strength and ability to reduce construction costs.

As seashells are readily available at seashores, organic waste is reduced, which in turn helps the environment.

Chapter 9
Geopolymer Concrete

9.1 Introduction

Geopolymer concrete is an inorganic polymer of high strength and lightweight that can be used in place of conventional concrete. Normal concrete uses ordinary Portland cement (OPC) as a binder, while geopolymer concrete uses a chemical and fly ash mixture as a binder. Geopolymer concrete has several advantages compared to conventional concrete.

Geopolymer concrete (GPC) is a hot topic in the construction industry throughout the world current days since it is a new and innovative material that is being used by many nations as a substitute to ordinary Portland concrete (OPC). The binding material used in geopolymer concrete is different; rather than using cement, waste products are being used, which makes the concrete unique.

Geopolymer concrete is made up of fine and coarse aggregates with fly ash as the main source and GGBFS as a fly ash substitute, as well as an alkaline activator.

The binding process in Portland cement concrete occurs when water interacts with the cement, producing CSH gel. Polymerization happens in the geopolymer concrete with an alkaline solution. The activator is responsible for binding all the materials together in a process similar to that of cement.

Generally, three factors are usually responsible for increasing the strength and durability properties of geopolymer concrete. They are

1. alkaline activator strength ratio
2. molarity of NaOH, and
3. impact of GGBS in fly ash.

The sodium silicate–sodium hydroxide (SS/SH) ratio to be mixed with the ingredients is normally 1.5–3.5, depending on the alkaline activator used and 6–18 M is used.

K. K. Poloju, *Advanced Materials and Sustainability in Civil Engineering*,
SpringerBriefs in Applied Sciences and Technology,
https://doi.org/10.1007/978-981-16-5949-2_9

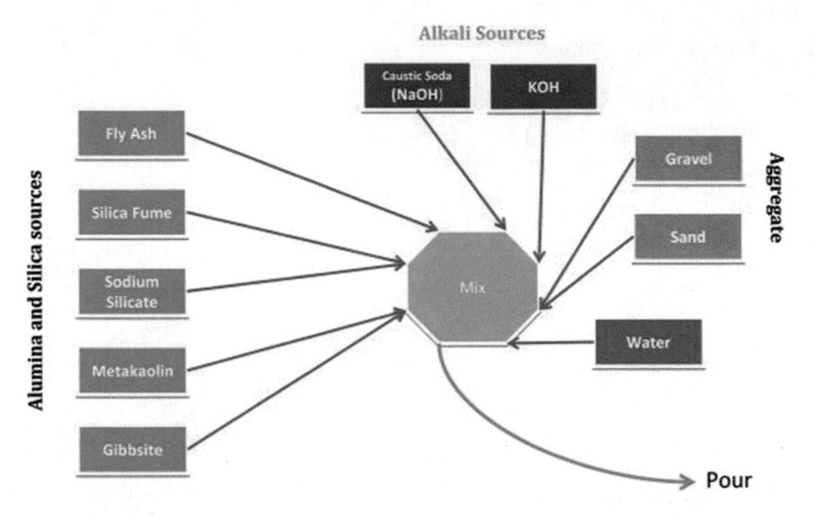

Fig. 9.1 Ingredients of GPC

9.2 Formation of Geopolymer Concrete

Geopolymer concrete is made up of several materials. The coarse and fine aggregates make up about 75–80% of the mass, equivalent to regular concrete. Crushed stone and gravel are examples of coarse aggregate, while sand is an example of fine aggregate. Chemical admixtures such as air entertainers and superplasticizers may be applied to the concrete mix to change its properties to suit the project's requirements. Superplasticizers minimize the amount of water in the mixture, making it much more workable. Air entrainers add air bubbles to the mixture. Geopolymer concrete mixtures, on the other hand, use a geopolymer paste to bond the aggregates together, rather than cement. Fly ash, which includes silicon and aluminium, is mixed with sodium hydroxide and sodium silicate solution to make geopolymer paste. The formation of GPC is shown in Fig. 9.1.

9.3 Ingredients Present in Geopolymer Concrete

9.3.1 Fly Ash

One of the most abundant materials on the planet is fly ash. Because of its role in the polymerization process, it is also an important component in the production of geopolymer concrete. Fly ash is a pozzolana that is powdery. Pozzolana is a substance that, when mixed with calcium hydroxide, has cementitious properties. Fly ash is the most common by-product of coal combustion in coal-fired power plants. Fly ash is divided into two "classes": Class F and Class C. Each type of fly ash has its own set of characteristics. The combustion of either anthracite or bituminous coal produces Class F fly ash.

This form of fly ash contains very little calcium oxide and has little or no self-cementing properties (also known as lime). To use Class F fly ash in concrete, it must be blended with a cementing agent, such as Portland cement, and an air-entraining admixture. Mixing it with ordinary concrete is not a very cost-effective operation. Class C fly ash, on the other hand, is formed when lignite or sub-bituminous coal is burned. It has self-cementing properties and a higher lime content than Class F fly ash, making it suitable for use in ordinary Portland cement-based concrete.

Fly ash of Class F is widely found in geopolymer concrete. Fly ash is made up of a variety of chemical compounds, including arsenic, lead, cadmium, mercury, and uranium, as well as other potentially harmful substances. When these chemicals undergo a polymerization reaction, however, their harmful properties are neutralized. Silicon dioxide and aluminium (III) oxide are the most important compounds in fly ash for geopolymer concrete. These are not only the main reactants in the polymerization reaction, but they also account for roughly 70% of the total weight of fly ash.

Currently, fly ash is recycled into ordinary Portland cement-based concrete at a rate of 1–1.5 pounds of fly ash per 1 pound of cement to maximize the overall strength of the concrete. The type of fly ash used is determined by the application. Class C fly ash, for example, should not be used in an atmosphere where it can come into contact with sulphate, as this form of fly ash can potentially reduce sulphate resistance. Class F fly ash, on the other hand, can improve sulphate resistance.

Ground-granulated blast furnace slag (GGBS)—Ground-granulated blast furnace slag is a waste or by-product of the steel or iron manufacturing process. It consists of adequate calcium and alumni content that would be helpful to support fly ash in achieving strength of geopolymer concrete.

- When compared to other construction materials, particularly ordinary Portland concrete, geopolymer concrete behaves considerably better.

Aggregates: The aggregate is primarily responsible for the mass of concrete. It occupies around 75–80% of the mass of GPC. However, the aggregate is of two types as shown in Fig. 9.2. They are as follows:

Fig. 9.2 Fine aggregate and coarse aggregate

Fine aggregate: The aggregate that can pass through a 4.75 mm sieve is known as fine aggregate.

Coarse aggregate: The aggregate retained on a 4.75 mm sieve is known as coarse aggregate.

Alkaline Activator: Alkaline activators are commonly made with two solutions (NaOH and SiO_2) as shown in Figs. 9.3 and 9.4. When these solutions are combined and mixed, they act as activators. The different types of molarity of sodium hydroxide have different effects on the strength of concrete. As the molarity increases, the amount of sodium hydroxide increases, which leads to a decrease in sodium silicate content. Pure NaOH pellets are mixed with water to obtain the solution of the required concentration. However, an activator is prepared with the addition of sodium silicate in the required quantity to the NaOH solution.

For example, the concentration of sodium hydroxide is taken as 8M and mixed with sodium silicate solution to prepare an alkaline activator solution. However, the amount of sodium silicate solution depends on the weight of the sodium hydroxide solution. Generally, the sodium silicate/ sodium hydroxide (SS/SH) is taken as 1, 1.5, 2, 2.5, or 3 according to the requirements.

Firstly, 320 g of NaOH (for 8 M) pellets are taken into 1 L of a jar and mixed with water and stirred completely to prepare the NaOH solution. The mixture is then weighed.

Later, sodium silicate solutions of adequate quantity (2.5 times the weight of sodium hydroxide solution) are added to the NaOH mixture to prepare an alkaline activator. The mixture should not be disturbed for 24 h before using it in the mix.

Fig. 9.3 Sodium hydroxide pellets

Fig. 9.4 Sodium silicate solution

9.4 Polymerization (Geopolymerization)

The formation of a geopolymer is defined by the exothermic process of polymerization. First, sodium hydroxide and sodium silicate solution dissolves the silicon and aluminium in fly ash. The dissolved ions then undergo a condensation reaction, resulting in the monomers –Si–O–Al–O, –Si–O–Al–O–Si–O, or –Si–O–Al–O–Si–O. The heat between 60 and 80 °C added to these monomers for 6–12 h causes them to polymerize or bind together and form rigid chains. The polymers are then allowed to harden.

Geopolymerization is an exothermic process that involves a chemical reaction between alumina silicates found in raw materials and alkaline activators to produce complex geopolymers.

Combinations of sodium hydroxide NaOH or potassium hydroxide KOH and sodium silicate Na_2SiO_3 or potassium silicate K_2SiO_3 have widely used activators.

In general, only sodium hydroxide or only potassium hydroxide can be used for activation. When only hydroxides are present, however, the reaction proceeds at a slower pace. As a result, hydroxide–silicate combinations (sodium hydroxide or potassium hydroxide and sodium silicate or potassium silicate) are more useful, as they speed up the reaction rate and polymerization process. In the dissolution of elements found in source materials, the concentration of sodium hydroxide is important. An increase in concentration causes more aluminosilicate to dissolve, resulting in a faster polymerization process.

9.5 Advantages of Geopolymer Concrete

The benefits of geopolymer concrete over traditional concrete are numerous. Concrete made with ordinary Portland cement is stronger, more resistant to chemicals and corrosion, and has a broader variety of applications.

Geopolymer concrete is not a magical substance; it is like ordinary concrete which needs to be specially tailored to be able to manage itself in unique circumstances. As opposed to ordinary concrete, geopolymer concrete is safer and more durable in extreme conditions.

9.6 High Strength Concrete

The compressive strength of geopolymer concrete is very high. In reality, geopolymer concrete had higher compressive strength than ordinary concrete. Geopolymer concrete is able to withstand over 10 MPa more than ordinary concrete after being given enough time to cure as per few research studies. Even so, the maximum compressive strength of ordinary concrete is significantly lower than that of geopolymer concrete. The lack of tensile strength in ordinary concrete is a disadvantage. As opposed to the extent of its compressive strength, its tensile strength is dangerously low. This means that concrete is brittle in conditions like earthquakes, where the material is subjected to several compressive and tensile forces. Geopolymer concrete, in addition to having high compressive strength, also has high tensile strength as opposed to ordinary concrete. While a higher tensile strength does not make geopolymer concrete earthquake proof, it is an advantage in certain cases.

9.7 Creep and Low Shrinkage

Shrinkage and creep are also very low in geopolymer concrete. Concrete shrinkage occurs as a result of the concrete drying and heating, as well as the evaporation of water from the concrete. Shrinkage allows the substance to crack, which can be harmful. Since geopolymer concrete does not hydrate like traditional concrete, it does not undergo nearly as much shrinkage. In reality, it shrinks 4/5 less than regular concrete. Geopolymer concrete also has a very low creep rate. Creep refers to a material's propensity to become permanently deformed as a result of repetitive forces acting on it. To construct a system that is both secure and long lasting, it is important to use a material that is unaffected by these major deformations as much as possible. Since these two influences do not affect geopolymer concrete, it has a significant advantage over ordinary concrete.

9.8 Resistance to Heat and Cold

Furthermore, geopolymer concrete has a high fire and heat resistance. It can withstand temperatures of over 1200 °C and remain stable (about 2200 °F.) Although no concrete is flammable, excessive heat (such as that generated by a fire) is extremely dangerous. It not only reduces the concrete's strength but also increases the risk of spalling. Spalling is the splitting of layers of concrete, which can be used as an "explosion". An "explosion" in the context of concrete refers to a loud popping sound rather than the act of dangerously launching concrete fragments at high speeds. Because of the high resistance of geopolymer concrete, the geopolymer concrete not only works well in high-heat settings but also prevents freezing. This is critical because thawing can cause damage to the concrete's structure. Concrete is a porous material, even though its pores are very small. Water seeps into the pores of the concrete and remains there until it evaporates. The water inside the concrete, on the other hand, can freeze and grow if the temperature drops below zero. After freezing, the water, now in the form of ice, would take up about 9% more space than it did when it is liquid. The concrete's framework will be destroyed as a result of this expansion, which will cause the concrete to spall. This mechanism, known as the freeze–thaw cycle, will continue to damage the concrete for the rest of its existence. Fly ash reduces the permeability of geopolymer concrete, making it less susceptible to water intrusion. Since the amount of water allowed to penetrate a concrete sample dictates the amount of damage caused by the freeze–thaw cycle, geopolymer concrete will not sustain as much damage as normal concrete.

9.9 Resistance to Chemicals

Chemical resistance is another advantage of geopolymer concrete. Acids, hazardous waste and saltwater have little effect on it. Geopolymer concrete, according to one report, has considerably higher acid resistance than ordinary concrete. Just about 5% of the compressive strength of geopolymer concrete made with Class F fly ash is lost after 8 weeks of immersion in sulphuric acid. This is due to the alkalinity of the water. Ordinary concrete, on the other hand, lost around 85% of its tensile strength in similar conditions.

Toxic and hazardous waste can also be contained in geopolymer concrete. Allowing toxic and hazardous wastes to be released into the atmosphere is harmful to both the environment and human health. Geopolymer concrete immobilizes radioactive wastes like uranium, trapping them and preventing them from escaping into the atmosphere. These hazardous wastes can be neutralized by mixing them with geopolymer concrete. Thc wastc is trapped inside as the mixture hardens into an impermeable solid. Geopolymer concrete, when used in this manner, may help to improve public health and safety.

Geopolymer concrete, unlike standard concrete, is also resistant to saltwater. Corrosion caused by saltwater is a significant issue in ordinary concrete. Saltwater continuously attacks and damages Portland cement-based concrete, creating cracks that allow more water to enter and cause more damage. While geopolymer concrete is affected by saltwater corrosion, it is not affected to the same degree as concrete.

This can be seen in the graph below, which plots time against mill amperes. A scientist tested the amount of corrosion caused by saltwater on samples of both geopolymer and ordinary concrete to show that geopolymer concrete is well suited to withstand saltwater applications. Water entered the concrete as saltwater corroded the materials. A direct current is generated between the water and the steel beams within the samples. When a higher amperage is reported in the graph, it means the concrete sample has been corroded more than when a lower amperage is recorded.

9.10 Geopolymer Concrete's Economic Viability

While geopolymer concrete is superior to ordinary Portland cement-based concrete in a variety of applications, it must also be a cost-effective alternative to replace the more conventional material in any application. It would not be a fair replacement if it proved to be too costly. As a result, both the financial commitments and the material's output must be addressed.

In terms of material costs, geopolymer concrete is considerably less expensive than Portland cement-based concrete. First and foremost, fly ash is a very low-cost material. In reality, since it is a waste product, it might be possible to obtain the material for free. Owing to the extremely high availability of fly ash, even though the vendor sells it, the price per ton will be slightly less than that of Portland cement. The cost of fly ash varies between 10 and 30% less than the cost of Portland cement.

Geopolymer concrete can also save money over its lifetime when compared to current concrete technology. Geopolymer concrete is much less vulnerable to damage than conventional concrete. Damaging factors such as spalling, freezing, and chemical contamination do not affect it, and therefore, it is highly durable compared to conventional concrete. Since this type of concrete has a longer lifetime, it would not be necessary to address the many issues that come with it regularly, saving vast sums of money that would otherwise be spent on maintenance costs.

9.11 Geopolymer Concrete Applications

Geopolymer concrete has a wide range of properties that allow it to be used in a variety of applications, giving it a significant advantage over traditional concrete. As it is used in a variety of applications, geopolymer concrete deserves more attention from researchers and the industry. It is worth noting that a single batch of geopolymer concrete will not necessarily have all the properties needed for each of the applications mentioned below.

Geopolymer concrete could be used in the design and maintenance of highways, bridges, and airport runways, for example. The US military is already using geopolymer concrete because of its ability to withstand the heat produced by aircraft taking off. According to one source, a runway made of geopolymer concrete can withstand the weight of a walking person after one hour, a car after four hours, and an Airbus or Boeing plane after only six hours due to its rapid strength benefit. A comparable runway made of standard cement-based concrete would have taken several days to reach the same strength level. Geopolymer concrete can be used in areas where a quick and reliable repair is needed, such as highways, because of its rapid strength gain property. The sooner a highway can be fixed, the sooner it can be reopened and traffic flow can be restored. Since geopolymer concrete has a high resistance to chloride, when calcium chloride road salts are used to lower the freezing point of water and keep roads open in the winter, it can experience less harm than roads and highways made of ordinary concrete.

Another area where geopolymer concrete could be used is in maritime environments. Geopolymer concrete can be used for concrete buildings that are constantly exposed to saltwater because of its high salt resistance. Piers, coastal bridges, and underwater concrete supports are only a couple of applications where geopolymer concrete can shine. Corrosion of steel supports found inside maritime concrete structures, such as bridges, is a major concern. Steel corrodes when it comes into contact with a material with a pH less than 11, and because seawater has a pH of 8, it can quickly corrode steel. To avoid this corrosion, the steel must be encased in concrete that is resistant to chloride corrosion. Geopolymer concrete may be used to keep seawater from touching the steel supports because of its resistance to chloride corrosion.

Geopolymer concrete is also better suited to cold climates than ordinary concrete. Geopolymer concrete, with its high resistance to freezing, can find widespread use in northern regions where concrete freezing is a common issue.

Geopolymer concrete can be used in a variety of highly acidic and hazardous conditions due to its high resistance to acids and toxic waste. Sewer pipes and landfills are examples of this type of setting. Geopolymer concrete may provide long-lasting structures that keep harmful contaminants out of the atmosphere. Corrosion prevention is particularly important in sewer pipes. Since the pipes are underground, repairing them is very costly and inefficient, particularly if the pipe is under a road, which would have to be partially rerouted.

Geopolymer concrete is a long-lasting construction material that outperforms ordinary concrete in many ways. Concrete made of geopolymer can be categorized as sustainable, but concrete made of ordinary Portland cement cannot. Furthermore, geopolymer concrete is made in a more efficient and energy-efficient manner. It is also more chemical, heat, cold, and corrosion resistant than traditional concrete. Since geopolymer concrete is still in its early stages of growth, its widespread adoption in the construction industry is yet to be seen. Geopolymers are appealing for use in high-cost, harsh environment applications because of their toughness. The sooner ordinary concrete is replaced with geopolymer concrete, the better for the climate and community. Hopefully, geopolymer concrete would soon overtake ordinary Portland cement as the most abundant man-made commodity on the planet.

Bibliography

1. Cement and concrete: environmental considerations from EBN. Environmental Building News, vol. 2, no. 2, March/April 1993
2. Eco smart concrete Seminar held in two cities in the United Arab Emirates—October 22, 2007 in Abu Dhabi and October 24, 2007 in Dubai
3. M.L. Gambhir, *Concrete Technology*, 3rd edn. (Tata McGraw Hill Publishers, New York, 2007)
4. S. Hemant, L.N. Mittal, P.D. Kulkarni, *Laboratory Manual on Concrete Technology* (CBS; 1st edition (January 1, 2010))
5. N. Krishna Raju, *Design of Concrete Mixes*, 5th edn. (CBS Publishers And Distributors Pvt Ltd, January 30, 2017)
6. V.M. Malhotra, Role of supplementary cementing materials in reducing greenhouse gas emissions, in *Concrete Technology for a Sustainable Development in the 21st Century* (CRC Press, London, 2000)
7. P.K. Mehta, *Greening of the Concrete Industry for Sustainable Development* (Concrete International, 2002)
8. A.R. Santhakumar, *Concrete Technology*, 3rd edn. (Oxford University Publishers, Oxford, 2009)
9. M.S. Shetty, *Concrete Technology*, first multi-color edn. (S. Chand Publications, 2005)
10. G.J. Venta, N. Bouzoubaa, B. Fournier, Production and use of supplementary cementing materials in canada and the resulting impact on greenhouse gas emissions reductions, in *Eighth CANMET/ACI International Conference on Fly Ash, Silica Fume, Slag and Natural Pozzolans in Concrete*, Supplementary Papers, Las Vegas, U.S.A., May 23–29, 2004, pp. 73–87
11. D. Patel, To study the properties of concrete as a replacement of cement with the marble dust powder. Int. J. Civ. Eng. Technol. **7**(4), 199–220 (2016)
12. J.N. Bhanushali, D.K. Mistry, A.M. Desai, M.M. Lad, R.D. Kumar, J.J. Patel, Scope of utilization of waste marble powder in concrete as partial substitution of cement. Int. Res. J. Eng. Technol. (IRJET) **05**(06) (2018). eISSN: 2395-0056
13. UN Environment Program, https://www.unenvironment.org/ (2019)
14. K.K. Poloju, A.Z.M. Al-Ruqaishi, M.S.H.A. Allamki, The advancement of ceramic waste in concrete. Int. J. Adv. Appl. Sci. **6**(11), 102–108 (2019)
15. K. Poloju, V. Anil, R.K. Manchiryal, Impact of nano silica on strength and durability properties of self-compacting concrete. Int. J. Adv. Appl. Sci. **4**(5), 120–126 (2017)
16. K. Katuwal, Comparative study of M35 concrete using marble dust as partial replacement of cement and fine aggregate. Int. J. Innov. Res. Sci. Eng. Technol. 66–78 (2017)
17. M. Trivedi, Partial replacement of cement with marble dust powder in cement concrete. Int. J. Res. Appl. Sci. Eng. Technol. (IJRASET) [Online] **5**(5) (2019)
18. R. Chandrakar, Cement replacement in concrete with marble dust powder. Int. Res. J. Eng. Technol. 1409–1417 (2017)

K. K. Poloju, *Advanced Materials and Sustainability in Civil Engineering*,
SpringerBriefs in Applied Sciences and Technology,
https://doi.org/10.1007/978-981-16-5949-2

19. A.A. Aliabdo, Reuse of waste marble dust in the production of cement and concrete. Constr. Build. Mater. 28–41 (2014)
20. M.K. Trivedi, Partial replacement of cement with marble dust powder in cement concrete. Int. J. Res. Appl. Sci. Eng. Technol. [Online] **5**(5), 73–85 (2017)
21. R. Kumar, S.K. Kumar (2015) Partial replacement of cement to concrete by marble dust powder. Int. J. Mod. Trends Sci. Technol. [Online] **2**(5) (2016). ISSN: 2455-3778
22. A.A.L. Yousef, O. Benjeddou, M.A. Khadimallah, A.M. Mohamed, C. Soussi, Study of the effects of marble powder amount on the self-compacting concretes properties by microstructure analysis on cement-marble powder pastes. Adv. Civ. Eng. **2018**(Article ID 6018613), 13 p (2018)
23. K. Kumar Poloju, C. Rahul, V. Anil, Glass fiber reinforced concrete (GFRC)—strength and stress strain behavior for different grades of concrete. Int. J. Eng. Technol. **7**(4.5), 707–712 (2018)
24. K.K. Poloju et al., Examine possible outcomes on strength properties for utilizing rubber waste on various grade of concrete. J. Adv. Res. Dyn. Control Syst. (08-Special Issue), 01–07 (2018)
25. F. Aydin, M. Saribiyik, Correlation between Schmidt Hammer and destructive compressions testing for concretes in existing buildings. Sci. Res. Essays [Online] **5**(13), 1644–1648 (2019)
26. ACIMAC, *World Production and Consumption of Ceramic Tiles*, 5th edn. (Association of Italian Manufacture of Machinery and Equipment for Ceramics, Baggiovara, Italy, 2017)
27. A. Agrawal, Use of ceramic waste in concrete: a review. Int. J. Eng. Res. **4**(3), 331–339 (2016)
28. A.V. Alves, T.F. Vieira, J. de Brito, J.R. Correia, Mechanical properties of structural concrete with fine recycled ceramic aggregates. Constr. Build. Mater. **64**, 103–113 (2014)
29. A. Camões, B. Aguiar, S. Jalali, Estimating compressive strength of concrete by mortar testing, in *INCOS 05 International Conference on Concrete for Structures*, Coimbra, Portugal (2005)
30. J.R. Correia, J. de Brito, A.S. Pereira, Effects on concrete durability of using recycled ceramic aggregates. Mater. Struct. **39**(2), 169–177 (2006)
31. E.A. Hansen, *Determination of the Tensile Strength of Concrete*, vol. 17 (Nordic Concrete Research-Publications, 1995), pp. 1–17
32. W. Jackiewicz-Rek, K. Załęgowski, A. Garbacz, B. Bissonnette, Properties of cement mortars modified with ceramic waste fillers. Procedia Eng. **108**, 681–687 (2015)
33. A.V.J. Meena, Experimental study of ceramic waste electric insulator powder used as a partial replacement of cement in concrete. J. Mater. Sci. Surf. Eng. **5**(4), 606–611 (2017)
34. H. Patel, N.K. Arora, S.R. Vaniya, The study of ceramic waste materials as partial replacement of cement: review. Int. J. Sci. Res. Dev. **3**(2), 863–865 (2015)
35. A.D. Raval, I.N. Patel, J. Pitroda, Ceramic waste: effective replacement of cement for establishing sustainable concrete. Int. J. Eng. Trends Technol. **4**(6), 2324–2329 (2013)
36. A.D. Raval, I.N. Patel, J. Pitroda, Re-use of ceramic industry wastes for the elaboration of eco-efficient concrete. Int. J. Adv. Eng. Res. Stud. **2**(3), 103–105 (2013)
37. A.D. Raval, I.N. Patel, J. Pitroda, Eco-efficient concretes: use of ceramic powder as a partial replacement of cement. Int. J. Innov. Technol. Exploring Eng. (IJITEE) **3**(2), 1–4 (2013)
38. R.M. Senthamarai, P.D. Manoharan, Concrete with ceramic waste aggregate. Cem. Concr. Compos. **27**(9–10), 910–913 (2005). https://doi.org/10.1016/j.cemconcomp.2005.04.003
39. A. Singh, V. Srivastava, Ceramic waste in concrete—a review, in *Proceedings of the IEEE International Conference Recent Advances on Engineering, Technology and Computational Sciences (RAETCS)*, Allahabad, India (2018), pp. 6–8
40. F.P. Torgal, S. Jalali, Compressive strength and durability properties of ceramic wastes based concrete. Mater. Struct. **44**(1), 155–167 (2011)
41. R. Yadav, G. Routiya, N. Jethwani, Effective replacement of cement for establishing sustainable concrete using ceramic waste. Int. J. Adv. Res. Sci. Eng. **6**(2), 75–80 (2017)
42. O. Zimbili, W. Salim, M. Ndambuki, A review on the usage of ceramic wastes in concrete production. Int. J. Civ. Environ. Struct. Constr. Architectural Eng. **8**(1), 91–95 (2014)
43. A.D. Raval, I.N. Patel, J. Pitroda, Eco-efficient concretes: use of ceramic powder as a partial replacement of cement. Int. J. Innov. Technol. Exploring Eng. (IJITEE) **3**(2) (2013)

44. A.R. Brough, A. Atkinson, Sodium silicate-based, alkali-activated slag mortars: part 1. Strength, hydration and microstructure. Cem. Concr. Res. **32**(6), 865–879 (2001)
45. A. Haidari, D. Tavakoli, A study on the mechanical properties of ground ceramic powder concrete incorporating nano-SiO_2 particles. Constr. Build. Mater. **38**, 255–264 (2012)
46. C. Sukesh, B.K. Katakam, P. Saha, K. Shyam Chamberlin, A study of sustainable industrial waste materials as partial replacement of cement. IPCSIT **28** (2012)
47. Y. Cheng, F. Huang, G.-L. Li, L. Xu, J. Hou, Test research on effect of ceramic polishing powder on carbonation and sulphate-corrosion resistance of concrete. Constr. Build. Mater. **55**, 440–446 (2014)
48. C. Medina Martinez, M.I. Guerra Romero, J.M. Moran del Pozo, J. Valdes, Use of ceramic waste in structural concretes. J. Student Res. ISSN: 2167-1907. www.jofsr.com
49. K.K. Poloju, R.K. Manchiryal, R. Chiranjeevi Rahul, Development of sustainable concrete by using paper industry waste. Elixir Civ. Eng. **102**, 44152–4415444152 (2017)
50. E. Vejmelkova, T. Kulovana, M. Keppert, P.K.M. Ondracek, M. Sedlmajer, R. Cerny, *Application of Waste Ceramics as Active Pozzolana in Concrete Production*, vol. 28 (IPCSIT, Singapore, 2012)
51. F. Pacheco-Torgal, S. Jalali, Reusing ceramic wastes in concrete. Constr. Build. Mater. **24**(5), 832–838 (2009)
52. A.R. Khaloo, Crushed tile coarse aggregate concrete. Cem. Concr. Aggregate **17**, 119–125 (1995)
53. V. Lopez, B. Llamas, A. Juan, J.M. Moran, I. Guerra, Eco-efficient concretes: impact of the use of white ceramic powder on the mechanical properties of concrete. Bio Syst. Eng. **96**, 559–564 (2007)
54. F.P. Torgal, S. Jalali, Reusing ceramic wastes in concrete. Constr. Build. Mater. **24**, 832–838 (2010)
55. C. Medina, M.I. Sanchez, M. Frias, Reuse of sanitary ceramic wastes as coarse aggregate in eco-efficient concretes. Cem. Concr. Compos. **34**, 48–54 (2012)
56. Z. Li, Y. Zhang, Development of sustainable cementitious materials. Hong Kong University of Science and Technology [Online]
57. K.K. Poloju, A. Shill, A.R. Zahid, Al. Balushi, S.R.S.A. Maawali, Determination of strength properties of concrete with marble powder. Int. J. Adv. Sci. Technol. **29**(08), 4004–4008 (2020)
58. K. Poloju, H. Darwesh, M. Tabanaj, Development of sustainable concrete using ceramic waste as partially replacement of cement. J. Student Res. (2017)
59. D. Hardjito, S. Wallah, D. Sumajouw, B. Rangan, On the development of fly ash-based geopolymer concrete. ACI Mater. J. **101**(6), 467–472 (2004)
60. S. Wallah, B. Rangan, Low-calcium fly ash-based geopolymer concrete: long-term properties. geopolymer.org (2006)
61. E. Ivan Diaz-Loya, E. Allouche, Engineering fly ash-based geopolymer concrete, in *2010 International Concrete Sustainability Conference* (2010)
62. Introduction: developments and applications in geopolymer. Geopolymer Institute. [Online Article]. Available: http://www.geopolymer.org/applications/introduction-developments-and-applications-in-geopolymer
63. D. Hardjito, B. Rangan, Development and properties of low-calcium fly ash-based geopolymer concrete. Curtin University of Technology (2005)
64. K.K. Poloju, K. Srinivasu, Impact of GGBS and strength ratio on mechanical properties of geopolymer concrete under ambient curing and oven curing. Mater. Today Proc. (2021). ISSN 2214-7853. https://doi.org/10.1016/j.matpr.2020.11.934
65. K.K. Poloju, C.R. Rollakanti, R.K. Manchiryal, A study on effect of alkaline activator on strength properties of geopolymer concrete. Int. J. Eng. Res. Manag. (IJERM) (2020). ISSN: 2349-2058. Special Issue, November 2020
66. How are seashells created? Or any other shell, such as a snail's or a turtle's? (2016)
67. M. Oliviaa, A.A. Mifshellaa, L. Darmayant, Mechanical properties of seashell concrete. Procedia Eng. **125**, 760–764 (2015)

68. P. Sasi Kumar, C. Suriya Kumar, P. Yuvaraj, B. Madhan Kumar, K. Jegan Mohan, A partial replacement for coarse aggregate by sea shell and cement by lime in concrete. Imperial J. Interdisc. Res. **2**(5), 1131–1136 (2016)
69. M. Mageswari, C.R. Manoj, M. Siddarthan, T.P. Saravanan, G. Princepatwa, To increase the strength of concrete by adding seashell as admixture. Int. J. Adv. Res. Civ. Struct. Environ. Infrastruct. Eng. Developing **2**(2), 165–174 (2016)
70. R. Yamuna Bharathi, S. Subhashini, T. Manvitha, S. Herald Lessly, Experimental study on partial replacement of coarse aggregate by seashell & partial replacement of cement by fly ash. Int. J. Latest Res. Eng. Technol. **2**(3), 69–76 (2016)
71. S. Vignesh, A partial replacement for coarse aggregate by seashell and cement by fly ash in concrete, in *National Conference on Research Advances in Communication, Computation, Electrical Science and Structures* (2015), pp. 28–33
72. H.G. Nahushananda Chakravarty, T. Mutusva, Investigation of properties of concrete with seashells as a coarse aggregate replacement in concrete. Int. J. Sci. Technol. **1**(1), 285–295 (2015)
73. M. Dixit, A. Jain, D. Kumawat, A. Swami, M. Sharma, Replacement of fine aggregate in concrete with municipal solid waste bottom ash from incinerator, in *5th International Conference of Euro Asia Civil Engineering Forum* (2016)
74. B.R. Etuk, I.F. Etuk, L.O. Asuquo, Feasibility of using sea shells ash as admixtures for concrete. J. Environ. Sci. Eng. A **1**, 121–127 (2012)
75. D.H. Nguyen, N. Sebaibi, M. Boutouil, L. Leleyter, F. Baraud, The use of seashell by-products in pervious concrete pavers, World Academy of Science, Engineering and Technology. Int. J. Civ. Environ. Eng. **7**(11) (2013)
76. K.K. Poloju, V. Anil, R.K. Manchiryal, Properties of concrete as influenced by shape and texture of fine aggregate. Am. J. Appl. Sci. Res. **2**(3), 61–69 (2017)
77. R. Kumar, Partial replacement of cement with marble dust powder. Int. J. Res. Appl. Sci. Eng. Technol. **5**(8), 106–114 (2015). ISSN: 2248-9622
78. D. Tavakolia, A. Heidari, M. Karimian, Properties of concretes produced with waste ceramic tile aggregate. Asian J. Civ. Eng. **14**, 369–382 (2013)

Printed in the United States
by Baker & Taylor Publisher Services